New PUZZLES IN LOGICAL DEDUCTION

by GEORGE J. SUMMERS

Dover Publications, Inc.

New York

Published in Canada by General Publishing Company, Ltd.,
30 Lesmill Road, Don Mills, Toronto, Ontario.
Published in the United Kingdom
by Constable and Company, Ltd.

New Puzzles in Logical Deduction is a new work,
first published in 1968 by Dover Publications, Inc.

Standard Book Number: 486-22089-3
Library of Congress Catalog Card Number: 68-19447

MANUFACTURED IN THE UNITED STATES OF AMERICA
Dover Publications, Inc.
180 Varick Street
New York, N.Y. 10014

Contents

An asterisk indicates a puzzle which requires an elementary knowledge of algebra.

Introduction

The puzzles in this book have been composed to resemble short "who done it" mysteries. Each puzzle contains a number of clues, and it is up to the reader, or "detective," to determine from these clues which of various solutions is correct (or, to continue the analogy, which of various suspects is the culprit). In some of the puzzles an actual criminal must be sought, but the greater part of them concern more-or-less innocent people, or numbers.

The general method for solving these puzzles is as follows: The question posed at the end of each puzzle states a condition that must be met by the solution. For example, "which one of the four teams—the Alleycats, the Bobcats, the Cougars, or the Domestics–won the pennant?" stipulates "won the pennant" as a condition. The clues also stipulate conditions, either explicit or implied, involving the various "suspects." What the "detective" must do is discover all the conditions, and then determine which one—and only one—"suspect" satisfies the condition stated in the question.

The puzzles are arranged in order of increasing difficulty, so if a reader begins at the beginning and works his way through the puzzles in order, he may find himself able to solve puzzles which were originally too hard for him. To aid the reader when he reaches an impasse there is a "hint"—printed upside down on the bottom of the page—designed to set him thinking in the right direction.

Of the fifty puzzles, thirty-six require no special knowledge. Numbers are involved in some of these, but no knowledge of algebra is necessary. The rules of play for THE HOSTESS, THE CLUB TRICK, and THE TENTH TRICK are based on those of Bridge, but you do not need to know how to play Bridge in order to solve these three puzzles.

Fourteen puzzles require knowledge of simple high school Algebra.

1

The Best Player

Mr. Scott, his sister, his son, and his daughter are tennis players. The following facts refer to the people mentioned:

1. The best player's twin and the worst player are of opposite sex.
2. The best player and the worst player are the same age.

Which one of the four is the best player?

HINT: How many people are the same age?

2

Middletown

Middletown includes the homes of four salesmen: Arden, Blair, Clyde, and Duane.

1. Each of the four homes is located at an intersection of two or more streets as indicated in the following map of their section of town:

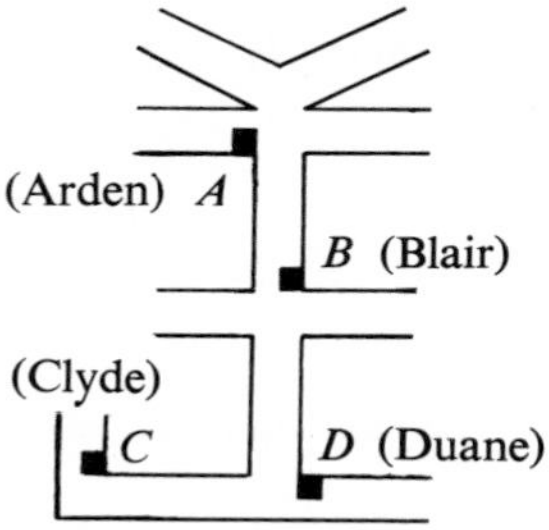

2. One day, simultaneously, Arden visited his friend Blair, Blair visited his friend Clyde, Clyde visited his friend Duane, and Duane visited his friend Arden.
3. On that day each salesman left his home and stopped at each house along each street in Middletown (each street contained houses along its entire length) before stopping at the home of his friend; but only one of the four men was able to do this without traveling along the same street more than once.

Which one of the four men traveled along all the streets in Middletown exactly once?

HINT: Is the number of passings through each intersection odd or even?

3

Murder in the Family

Murder occurred one evening in the home of a married couple and their son and daughter. One member of the family murdered another member, the third member witnessed the crime, and the fourth member was an accessory after the fact.

1. The accessory and the witness were of opposite sex.
2. The oldest member and the witness were of opposite sex.
3. The youngest member and the victim were of opposite sex.
4. The accessory was older than the victim.
5. The father was the oldest member.
6. The killer was not the youngest member.

Which one of the four—father, mother, son, or daughter—was the killer?

HINT: Which role did the youngest member play? Which member of the family was the youngest member?

4

Three A's

In the multiplication problem below, each letter represents a different digit:

$$
\begin{array}{r}
A\ S \\
\times\quad A \\
\hline
M\ A\ N
\end{array}
$$

Which one of the ten digits does A represent?

HINT: If there is a solution (and there is), only the value of A need be found.

5

Mary's Ideal Man

Mary's ideal man is tall, dark, and handsome. She knows four men: Alec, Bill, Carl, and Dave. Only one of the four men has all of the characteristics Mary requires.

1. Only three of the men are tall, only two are dark, and only one is handsome.
2. Each of the four men has at least one of the required traits.
3. Alec and Bill have the same complexion.
4. Bill and Carl are the same height.
5. Carl and Dave are not both tall.

Which one of the four men satisfies all of Mary's requirements?

HINT: How many men are both tall and dark?

6

Everybody Lied

When a psychiatrist was found murdered in his apartment, four of his patients were questioned about his death.

I. The police knew from the testimony of witnesses that each of the four patients had been alone with the psychiatrist in his apartment just once on the day of his death.

II. Before the four patients were questioned they met and agreed that every statement each patient made to the police would be a lie.

Each of the patients made two statements, as follows:

AVERY: 1. None of us four killed the psychiatrist.
2. The psychiatrist was alive when I left.

BLAKE: 3. I was the second to arrive.
4. The psychiatrist was dead when I arrived.

CROWN: 5. I was the third to arrive.
6. The psychiatrist was alive when I left.

DAVIS: 7. The killer did not arrive after I did.
8. The psychiatrist was dead when I arrived.

Which one of the four patients killed the psychiatrist?

HINT: Denying each of the eight statements leads to a determination of the order in which the men arrived and of when the psychiatrist was killed.

7

The Triangular Pen

A farmer built a triangular pen for his chickens. The pen was made of a wire mesh attached to posts imbedded in the ground.

1. The posts were spaced at equal intervals along each side of the pen.
2. The wire mesh, of uniform width, was attached to the posts at equal heights above the ground.
3. The farmer made the following entry in a notebook:

Cost of wire mesh for side of pen facing barn: $10
Cost of wire mesh for side of pen facing pond: $20
Cost of wire mesh for side of pen facing home: $30

4. He paid for the wire mesh with only ten-dollar bills, and received no change.
5. He paid with a different number of ten-dollar bills for the wire mesh along each side of the pen.
6. Exactly one of the three costs in his entry was incorrect.

Which one of the three costs was incorrect?

HINT: The costs of the sides of the pen must be in the same ratio as their lengths. What relative lengths of the sides of the pen are possible?

8

Speaking of Bets

"The three of us made some bets.

1. First, *A* won from *B* as much as *A* had originally.
2. Next, *B* won from *C* as much as *B* then had left.
3. Finally, *C* won from *A* as much as *C* then had left.
4. We ended up having equal amounts of money.
5. I began with 50 cents."

Which one of the three—A, B, or C—is the speaker?

HINT: Let *a*, *b*, and *c* be the amounts of money *A*, *B*, and *C* had, respectively, before the betting. Then represent algebraically the amounts of money each had after the betting. Only one of the three people could have begun with 50¢.

9

The Trump Suit

In a certain card game, one of the hands dealt contains:

1. Exactly thirteen cards.
2. At least one card in each suit.
3. A different number of cards in each suit.
4. A total of five hearts and diamonds.
5. A total of six hearts and spades.
6. Exactly two cards in the "trump" suit.

Which one of the four suits—hearts, spades, diamonds, or clubs—is the "trump" suit?

HINT: How many hearts are there?

10

Malice and Alice

Alice, Alice's husband, their son, their daughter, and Alice's brother were involved in a murder. One of the five killed one of the other four. The following facts refer to the five people mentioned:

1. A man and a woman were together in a bar at the time of the murder.
2. The victim and the killer were together on a beach at the time of the murder.
3. One of the two children was alone at the time of the murder.
4. Alice and her husband were not together at the time of the murder.
5. The victim's twin was innocent.
6. The killer was younger than the victim.

Which one of the five was the victim?

HINT: Where was Alice and who was with her?

11

A Week in Arlington

In the town of Arlington the supermarket, the department store, and the bank are open together on one day each week.

1. Each of the three places is open four days a week.
2. On Sunday all three places are closed.
3. None of the three places is open on three consecutive days.
4. On six consecutive days:
   the department store was closed on the first day
   the supermarket was closed on the second day
   the bank was closed on the third day
   the supermarket was closed on the fourth day
   the department store was closed on the fifth day
   the bank was closed on the sixth day

On which one of the seven days are all three places in Arlington open?

HINT: Only one day can be the first of the six consecutive days mentioned; otherwise a contradiction arises.

12

Eunice's Marital Status

At a party Jack saw Eunice standing alone at the punch bowl.

1. There were nineteen people altogether at the party.
2. Each of seven people came alone; each of the rest came with a member of the opposite sex.
3. The couples who came to the party were either engaged to each other or married to each other.
4. The women who came alone were unattached.
5. No man who came alone was engaged.
6. The number of engaged men present equalled the number of married men present.
7. The number of married men who came alone equalled the number of unattached men who came alone.
8. Of the married women, engaged women, and unattached women present, Eunice belonged to the largest group.
9. Jack, who was unattached, wanted to know which group of women Eunice belonged to.

Which one of the three groups of women did Eunice belong to?

HINT: The three numbers indicating the sizes of the three groups can be represented in terms of the same letter. By making an inference from one of the statements and applying it to the algebraic representation of one number, you can restrict the value of the letter to one number.

13

A Smart Man

Of Aaron, Brian, and Colin, only one man is smart.

Aaron says truthfully:

1. If I am not smart, I will not pass Physics.
2. If I am smart, I will pass Chemistry.

Brian says truthfully:

3. If I am not smart, I will not pass Chemistry.
4. If I am smart, I will pass Physics.

Colin says truthfully:

5. If I am not smart, I will not pass Physics.
6. If I am smart, I will pass Physics.

While

I. the smart man is the only man to pass one particular subject,
II. the smart man is also the only man to fail the other particular subject.

Which one of the three men is smart?

HINT: Can any man pass more than one subject? Which subject was passed by the smart man?

14

The Murderer

Three suspects, named Adam, Brad, and Cole, were questioned at different times about the murder of Dale.

Each of the following three statements was made by one of the three men:

1. Adam is innocent.
2. Brad is telling the truth.
3. Cole is lying.

Statement [1] was made first; but statements [2] and [3] are not necessarily in temporal order, though each refers to a statement made earlier.

I. Each man made one of these statements, referring to another of the suspects.

II. The murderer, who was one of the three men, made a false statement.

Which one of the three men was the murderer?

HINT: Who could have made each statement? The truth or falsity of each statement depends upon whether Adam is innocent or guilty. If Adam is guilty, he made a statement implying he was innocent. If Brad or Cole is guilty, one of them made a statement implying Adam was guilty.

15

The Missing Digit

In the addition problem below, each letter represents a different digit:

$$\begin{array}{r} A\ B \\ C\ D \\ E\ F \\ +\ G\ H \\ \hline I\ I\ I \end{array}$$

Which one of the ten digits is missing?

HINT: What digit must I represent?

16

One, Two, or Four

A coin game requires:

1. Ten coins in one pile.
2. That each player take one, two, or four coins from the pile at alternate turns.
3. That the player who takes the last coin loses.

I. When Austin and Brooks play, Austin goes first and Brooks goes second.

II. Each player always makes a move that allows him to win, if possible; if there is no way for him to win, then he always makes a move that allows a tie if possible.

Must one of the two men win? If so, which one?

HINT: It must first be decided whether drawing from one coin is a winning, losing, or tieing position; then whether drawing from two coins is a winning, losing, or tieing position; and so on through drawing from ten coins.

17

The Student Thief

Professor Dimwit's answer key to a Physics test was stolen one day during one of his Physics classes. Only three students—Amos, Burt, and Cobb—had the opportunity to steal the answer key.

1. Five Physics classes had been held in the room that day.
2. Amos attended only two of the classes.
3. Burt attended only three of the classes.
4. Cobb attended only four of the classes.
5. The professor conducted only three of the classes.
6. Each of the three students attended only two of the professor's classes.
7. No two of the five classes were attended by the same group of students from the three students under suspicion.
8. Two of the three students, who attended one of the professor's classes that the third student did not attend, were proven innocent of the theft.

Which one of the three students stole the answer key?

HINT: How many classes not conducted by the professor did each student attend? To satisfy [6], must each of his classes have been attended by two of the three students?

18

The Four Groves

Mr. Sloan has four groves: an apple grove, a lemon grove, an orange grove, and a peach grove.

1. Each grove contains rows of trees with the same number of trees in each row.
2. The apple grove has the least number of rows of trees, the lemon grove has one more row of trees than the apple grove, the orange grove has one more row of trees than the lemon grove, and the peach grove has one more row of trees than the orange grove.
3. The number of trees in the interior of each of three groves equals the number of trees on its border.

In which one of the four groves does the number of trees in the interior not equal the number of trees on the border?

HINT: Represent algebraically the number of trees in the interior and the number of trees on the border of the three groves mentioned in statement 3. There are only four possible pairs of values for the numbers of trees along adjacent sides of these three groves.

19

Hubert's First Watch

Hubert was one of a group of men hired by a jewelry company as an early-morning watchman.

1. For no more than 100 days Hubert was on a rotating system of standing watch.
2. Hubert's first and last watches were the only ones of his to occur on a Sunday.
3. Hubert's first and last watches occurred on the same date of different months.
4. The months in which Hubert's first and last watches occurred had the same number of days.

In which one of the twelve months did Hubert have his first watch?

HINT: The number of days occurring between Hubert's first and last watches must be a multiple of what number? This number of days must also be equal to the number of days in one, two, or three months.

20

The Square Table

Part I

Alden, Brent, Clark, and Doyle were seated around a square table in a restaurant when Doyle fell dead from poison. When questioned by a detective each man made two statements as follows:

ALDEN: 1. I sat next to Brent.
2. Brent or Clark sat on my right and that person could not have poisoned Doyle.

BRENT: 3. I sat next to Clark.
4. Alden or Clark sat on Doyle's right and that person could not have poisoned Doyle.

CLARK: 5. I sat across from Doyle.
6. If only one of us is lying that person poisoned Doyle.

After talking to the waiter who had served them, the detective told them, truthfully:

7. Only one of you lied.
8. One of you poisoned Doyle.

Which one of the three men poisoned Doyle?

HINT: What one arrangement of the four positions allows only one person to have lied?

Part II

The wives of Alden, Brent, and Clark witnessed Doyle's death by poison when the four men were seated around the square table in the restaurant. When questioned by the detective each woman made two statements, referring to the suspects by their first names, as follows:

RAY'S WIFE: 1. Ray sat next to Sid.
2. Sid or Ted sat on Ray's right and he could not have poisoned Doyle.

SID'S WIFE: 3. Sid sat next to Ted.
4. Ray or Ted sat on Doyle's right and he could not have poisoned Doyle.

TED'S WIFE: 5. Ted sat next to Doyle.
6. If only one of us is lying that person is the murderer's wife.

After talking to the waiter who had served the four men, the detective told the women, truthfully:

7. Only one of you lied.

Which one of the three women was the murderer's wife?

NOTE: *The solution to Part II must be consistent with the statements in Part I.*

HINT: What one arrangement of the four positions allows only one person to have lied and is compatible with the arrangement in Part I?

21

The Two Brothers

Albert, Barney, Curtis, Dwight, Emmett, and Farley are art collectors, two of whom are brothers. One day, when they were all at an art fair, the men bought art objects as described below:

1. The price of each art object was a whole number of cents.
2. Albert bought 1 art object, Barney bought 2, Curtis bought 3, Dwight bought 4, Emmett bought 5, and Farley bought 6.
3. The two brothers paid the same amount for each of the art objects they bought.
4. Each of the other four men paid twice as much for each object they bought as the two brothers paid for each of their objects.
5. Altogether the six men spent $1000 for the objects they bought.

Which two of the six men are brothers?

HINT: What was the number of art objects bought by each pair of men? What was the sum of the number of art objects bought by each pair of men plus twice the number of art objects bought by the other four men. One of these sums must divide 100,000 exactly.

22

One, Three, or Four

A coin game requires:

1. Nine coins in one pile.
2. That each player take one, three, or four coins from the pile at alternate turns.
3. That the player who takes the last coin wins.

I. When Aubrey and Blaine play, Aubrey goes first and Blaine goes second.

II. Each player always makes a move that allows him to win, if possible; if there is no way for him to win, then he always makes a move that allows a tie, if possible.

Must one of the two men win? If so, which one?

HINT: It must first be decided whether drawing from one coin is a winning, losing, or tieing position; then whether drawing from two coins is a winning, losing, or tieing position; and so on through drawing from nine coins.

23

A Timely Death

One evening five explorers named Wilson, Xavier, Yeoman, Zenger, and Osborn made separate camps along the banks of a river.

Wilson communicated with the other men by radio at various times during the night. When he received no reply from Osborn after 10:30 that night, Wilson communicated with the other three men to express his concern.

The next morning Osborn was found dead; he had been murdered. Evidence at the scene of the crime indicated that the killer had approached Osborn's camp by boat from the river. Each of the men had had access to a canoe on the previous night.

Wilson suspected that either Xavier, Yeoman, or Zenger had killed Osborn. From the following facts Wilson was able to eliminate two of these men as suspects:

1. Osborn was killed in his camp before 10:30 on the previous night; he had been shot and had died instantly.
2. The killer traveled to Osborn's camp and returned to his own camp by canoe.
3. Xavier's camp was located directly downstream from Osborn's camp, Yeoman's camp was located directly across the river from Osborn's camp, and Zenger's camp was located directly upstream from Osborn's camp.
4. It would require at least 80 minutes for each of the three men to get to Osborn's camp and to return to his own camp by canoe.
5. There was a strong current in the river.

6. Wilson received replies to his radio calls at the following times:

FROM	AT
Xavier	8:15
Yeoman	8:20
Zenger	8:25
Osborn	9:15
Xavier	9:40
Yeoman	9:45
Zenger	9:50
Xavier	10:55
Yeoman	11:00
Zenger	11:05

Which one of the three men could Wilson not eliminate as a suspect?

HINT: What was the latest time at which Osborn could have been killed? Which two men could not have had time to kill Osborn and answer both calls?

24

Turnabout

PROBLEM I	PROBLEM II	PROBLEM III
$A\ R\ B$	$A\ R\ S\ B$	$A\ R\ S\ T\ B$
$\times\ \ C$	$\times\ \ \ C$	$\times\ \ \ \ C$

1. In each of the three multiplication problems above, each letter represents a different digit; however, each letter does not necessarily represent the same digit in one problem as it does in another problem.
2. In each of two of these multiplication problems, the digits in the product are the same as the digits in the number multiplied by *C*, except that the order is reversed.

In which one of the three multiplication problems are the digits in the product not the same as the digits in the number multiplied by C?

HINT: In the two problems mentioned in statement 2, $C \times A$ must be less than or equal to *B*, and $C \times B$ must end in *A*.

25

A Week in Burmingham

The town of Burmingham has a supermarket, a department store, and a bank. On the day I went into Burmingham, the bank was open.

1. The supermarket, the department store, and the bank are not open together on any day of the week.
2. The department store is open four days a week.
3. The supermarket is open five days a week.
4. All three places are closed on Sunday and Wednesday.
5. On three consecutive days:
   the department store was closed on the first day
   the bank was closed on the second day
   the supermarket was closed on the third day
6. On three consecutive days:
   the bank was closed on the first day
   the supermarket was closed on the second day
   the department store was closed on the third day

On which one of the seven days did I go into the town of Burmingham?

HINT: The beginning day of each sequence can be determined from a consideration of the supermarket's schedule.

26

Cards on the Table

Eight numbered cards lie face down on a table in the relative positions shown in the diagram below.

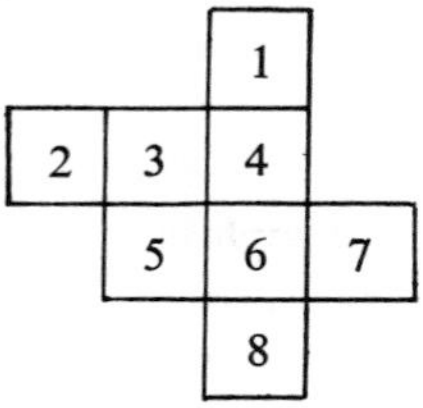

Of the eight cards:

1. There is at least one Queen.
2. Every Queen lies between two Kings.
3. There is at least one King between two Jacks.
4. No Jack borders on a Queen.
5. There is exactly one Ace.
6. No King borders on the Ace.
7. At least one King borders on a King.
8. Each card is either a King, a Queen, a Jack, or an Ace.

Which one of the eight numbered cards is the Ace?

Hint: Which numbered cards can be Queens?

27

The Murderess

Three women, named Anna, Babs, and Cora, were questioned about the murder of Dana. One of the three women committed the murder, the second was an accomplice in the murder, and the third was innocent of any involvement in the murder.

Each of the following three statements was made by one of the three women:

1. Anna is not the accomplice.
2. Babs is not the murderess.
3. Cora is not the innocent one.

I. Each statement refers to a woman other than the speaker.
II. The innocent woman made at least one of these statements.
III. Only the innocent woman told the truth.

Which one of the three women was the murderess?

HINT: How many statements were made by the innocent one?

28

Conversation Stopper

Amelia, Brenda, Cheryl, and Denise went to a party.

1. By 8 P.M., Amelia and her husband having already arrived at the party, there were no more than 100 people, all conversing in groups of five people each.
2. By 9 P.M., only Brenda and her husband having arrived at the party after 8 P.M., the people were conversing in groups of four people each.
3. By 10 P.M., only Cheryl and her husband having arrived at the party after 9 P.M., the people were conversing in groups of three people each.
4. By 11 P.M., only Denise and her husband having arrived after 10 P.M., the people were conversing in groups of two people each.
5. One of the four women mentioned, who suspected her husband of infidelity, had planned to let her husband go to the party without her, then to arrive at the party one hour later; but she changed her mind.
6. If the woman who suspected her husband of infidelity had carried out her plan, it would have been impossible for the people, including her husband but not herself, to converse in smaller groups containing equal numbers of people at one of the four times mentioned.

Which one of the four women suspected her husband of infidelity?

HINT: Represent algebraically the number of people present at each hour and use the fact that two people arrived at each of three intervals to form three equations.

29

The Constant Winner

Abe, Ben, Cal, and Don played some games in which they took turns drawing chips from a pile. The same man won every game.

1. The four men played fifty games with an even number of chips in a pile at the beginning of each game: two chips in a pile at the beginning of the first game, four chips in a pile at the beginning of the second game, and so on through 100 chips in a pile at the beginning of the fiftieth game.
2. During the entire series of fifty games, each of the four men always took the same number of chips: he always took one chip or he always took two chips; if only one chip remained and a player had decided to always draw two chips, he "passed" and the turn went to the next player.
3. The order in which they drew chips from a pile was always the same: first Abe, then Ben, then Cal, then Don.
4. The player who took the last chip was the winner in each game.

Which one of the four men won every game?

HINT: From [2], what are the possible four-way combinations for taking chips from a pile? Which combination always results in the same winner, no matter how many even chips are in a pile?

30

The Woman Hector Will Marry

Hector has been dating three women: Annette, Bernice, and Claudia.

Annette says truthfully:

1. If I talk a lot, then Bernice talks a lot.
2. If I am stubborn, then Claudia is stubborn.

Bernice says truthfully:

3. If I nag, then Claudia nags.
4. If I talk a lot, then Annette talks a lot.

Claudia says truthfully:

5. If I nag, then Annette nags.
6. If Bernice is stubborn, then I am not stubborn.

Hector says truthfully:

7. Each of the three traits is possessed by at least one of the three women.
8. Two of the women possess the same unfortunate traits.
9. Of the three women, the woman I will marry has only one of the three unfortunate traits.

Which one of the three women will Hector marry?

HINT: What are the possible combinations of women possessing each of three traits? There are three combinations for each trait. Who are the two women Hector will not marry?

31

Father and Son

Arnold, Barton, Claude, and Dennis are stock brokers, one man of whom is the father of one of the other three men. One day when they were all at the stock exchange, the men bought shares as described below.

1. Arnold bought only shares at $3 each, Barton bought only shares at $4 each, Claude bought only shares at $6 each, and Dennis bought only shares at $8 each.
2. The father bought the greatest number of shares, paying $72 altogether.
3. The son bought the least number of shares, paying $24 altogether.
4. The total money spent by the four men for their combined shares was $161.

Which one of the four men is the father? Which one of the four men is his son?

HINT: Form an equation from statements 1 and 4. If each man was the father, or the son, how many shares did he buy? If a number divides exactly four of the five terms in an equation then that number must divide exactly into the fifth term.

32

The Race

Alan, Bart, Clay, and Dick competed in a race where each man finished in a different position. The four men, notorious liars, reported the results of the race as follows:

ALAN: 1. I came in just before Bart.
2. I did not come in first.
BART: 3. I came in just before Clay.
4. I did not come in second.
CLAY: 5. I came in just before Dick.
6. I did not come in third.
DICK: 7. I came in just before Alan.
8. I did not come in last.

I. Only two of the above statements were true.
II. The man who won the race made at least one true statement.

Which one of the four men won the race?

HINT: Considered together, a certain number of statements 1, 3, 5, and 7 cannot be false. Considered together, a certain number of statements 2, 4, 6, and 8 cannot be false.

33

Murder by Profession

Bell and Cass were Alex White's sisters; Dean and Earl were Faye Black's brothers. (Alex was a man and Faye was a woman.) Their occupations were as listed below:

	Alex – doctor		Dean – doctor
White's	Bell – doctor	Black's	Earl – lawyer
	Cass – lawyer		Faye – lawyer

One night while two of these people were in a bar, two were on a beach, and two were at a movie, one of the two people on the beach killed the other.

The following facts refer to the people mentioned above:

1. A doctor and a lawyer were in the bar.
2. The two people at the movie had the same occupation.
3. The victim and the killer were twins.

4a. The victim was married to one of the two persons in the bar.

4b. The killer was married to the other person in the bar.

5. The victim and the victim's spouse had different occupations.

6a. One of the two persons at the movie was the ex-spouse of one of the two persons in the bar.

6b. The other person at the movie and the doctor in the bar were former roommates (same sex).

Which one of the six was the killer?

HINT: What was the relationship between the two people in the bar? What was the sex of each of the two people in the bar?

34

Six G's

In the multiplication problem below, each letter represents a different digit:

$$
\begin{array}{r}
A\ B\ C\ D\ E \\
\times \qquad\quad F \\
\hline
G\ G\ G\ G\ G\ G
\end{array}
$$

Which one of the ten digits does G represent?

HINT: What are the possible exact divisors of $G \times 111111$? Is G a multiple of F? Which digit does F represent?

35

Two or Three

A coin game requires:

1. Twelve coins in one pile.
2. That each player take two or three coins from the pile at alternate turns.
3. That the player who takes the last coin loses.

I. When Armand and Buford play, Armand goes first and Buford goes second.

II. Each player always makes a move that allows him to win, if possible; if there is no way for him to win, then he always makes a move that allows a tie, if possible.

Must one of the two men win? If so, which one?

HINT: It must first be decided whether drawing from one coin is a winning, losing, or tieing position; then whether drawing from two coins is a winning, losing, or tieing position; and so on through drawing from twelve coins.

36

A Week in Cantonville

In the town of Cantonville the supermarket, the department store, and the bank are open together on only one day each week.

1. The supermarket is open five days a week.
2. The department store is open four days a week.
3. The bank is open three days a week.
4. On three consecutive days:
   the bank was closed on the first day
   the department store was closed on the second day
   the supermarket was closed on the third day
5. On Sunday all three places are closed.
6. The bank is not open on two consecutive days.
7. The department store is not open on three consecutive days.
8. The supermarket is not open on four consecutive days.
9. All three places are not open on Saturday and Monday.

On which one of the seven days are all three places in Cantonville open?

HINT: What places are open on Saturday? What places are open on Monday?

37

The Book Shelf

When Mrs. Drake, the head librarian, asked each of her three assistants how many books would fill a certain shelf, she received the following replies:

MRS. ASTOR—2 catalogues, 3 dictionaries, and 3 encyclopedias will exactly fill this shelf

MRS. BRICE—4 catalogues, 3 dictionaries, and 2 encyclopedias will exactly fill this shelf

MRS. CRANE—4 catalogues, 4 dictionaries, and 3 encyclopedias will exactly fill this shelf

1. Only two of the assistants were correct in their replies.
2. Attempting to fill the shelf with books of the same type, Mrs. Drake discovered that only one type would exactly fill the shelf.
3. Mrs. Drake required 15 books of this one type to exactly fill the shelf.
4. All the catalogues were the same size, all the dictionaries were the same size, and all the encyclopedias were the same size.

Assuming that the books were of such widths as to make negligible the widths of the spaces between pairs of books, ***with which one of the three types of books did Mrs. Drake exactly fill the shelf?***

HINT: Represent algebraically the width of the bookcase and the combined widths of fifteen books of each of the three kinds. The three statements made by the women yield three pairs of equations involving the width of the bookcase and the combined widths of three kinds of books; only one pair of equations allows a positive value for the width of each kind of book. The width of the kind of book that can fill the shelf must not be a multiple of the width of either of the other two kinds of books.

38

The Hostess

Four women were playing a card game in which (a) a player must play a card in the suit led, if possible, at each trick (otherwise, a player may play any card); (b) a player who wins a trick must lead at the next trick. Ten tricks had already been played and there were three more tricks left to be played.

1. At trick number eleven: Alma led a club; Bess played a diamond, Cleo played a heart, and Dina played a spade, not necessarily in that order.
2. The hostess won the twelfth trick and led a heart at trick number thirteen.
3. A different person led at each of the last three tricks.
4. All four suits were played at each of the last three tricks, a "trump" winning each trick. (A trump is any card in a certain suit that may be (a) played when a player has no cards in the suit led—in this event a card in the trump suit beats all cards in the other three suits; or (b) led, as any other suit may be led.)
5. A different person won each of the last three tricks.
6. The partner of the hostess held three red cards.

Which one of the four women was the hostess?

HINT: Which suit was trump? Who played a trump at trick number twelve?

39

The Rectangular Table

Harry and his wife, Harriet, gave a dinner party to which they invited: his brother, Barry, and Barry's wife, Barbara; his sister, Samantha, and Samantha's husband, Samuel; and his neighbor, Nathan, and Nathan's wife, Natalie. While they were all seated at the table, one person produced a gun and shot another person. The chairs were arranged around the table as in the diagram below:

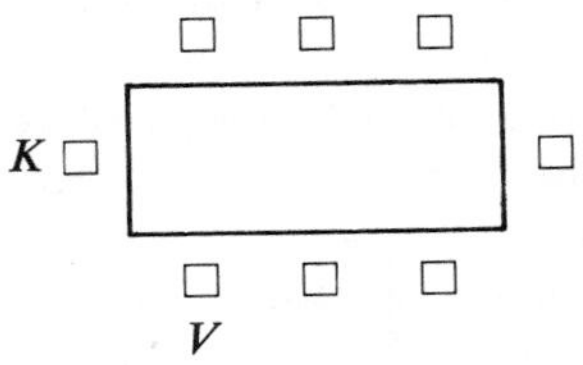

1. The killer sat in the chair marked *K*.
2. The victim sat in the chair marked *V*.
3. Every man sat opposite his wife.
4. The host was the only man who sat between two women (i.e., to the left of one and to the right of another, around the perimeter of the table).
5. The host did not sit next to his sister.
6. The hostess did not sit next to the host's brother.
7. The victim was the killer's former spouse.

Which one of the eight was the killer?

HINT: How were men and women seated in relation to each other? Note that Barry and Samantha are brother and sister.

40

Deanna's Sister

Deanna went shopping with her mother to buy some candy and favors for her sister's birthday party. Deanna's mother was to buy the favors, while Deanna was to buy the candy. The number of candies bought and the number of favors bought, together with the amount of money spent, are described below:

1. Deanna had with her thirteen coins, consisting of only three denominations: pennies, nickels, and quarters; she spent them all on the candy.
2. The candy she bought for Althea cost 2¢ each, the candy she bought for Blythe cost 3¢ each, and the candy she bought for Carrie cost 6¢ each.
3. She bought a different number of candies for each of the three girls, and she bought more than one candy for each girl.
4. For two kinds of candy she spent equal amounts of money.
5. Her mother bought a number of favors, each favor selling individually for the same amount of money; she paid $4.80 for the favors.
6. The number of favors bought was equal to the number of candies bought.
7. The girl for whom Deanna bought the greatest number of candies was her sister.

Which one of the three girls was Deanna's sister?

HINT: Form five equations from statements 1, 2, 5, and 6. Of three possible equations formed from statement 4, only one is correct. Whether the sums and products of various quantities in the equations are odd or even should be considered.

41

The Cube

Three views of the same cube are shown below:

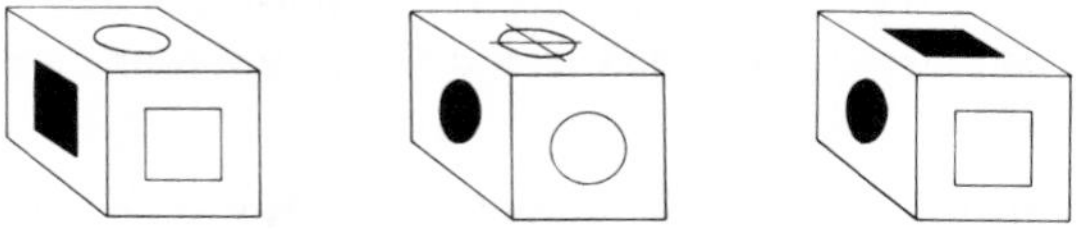

VIEW 1 VIEW 2 VIEW 3

As can be seen, there is one of five different figures on each face shown in these views: 

A little analysis indicates that one of these five figures must occur twice on the cube. In fact, any one of three figures can occur twice.

However, the owner of the cube states truthfully: "In each of the three views the figure that occurs twice is not on the bottom face of the cube." Now, only one figure can occur twice.

Which one of the five figures occurs twice on the cube?

NOTE: *If you find it difficult to visualize the cube's six faces, you might make a cube out of paper or draw a multiview cube as shown below. The bottom face will be the only face not seen.*

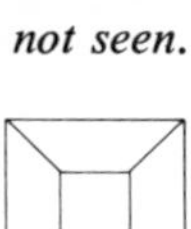

HINT: Either a given figure occurs once or it occurs twice. If it occurs once then the same face (containing the figure) appears in two views. If it occurs twice then the same face for each of three other figures appears in two views.

42

The Club Trick

Four women were playing a card game in which (a) a player must play a card in the suit led, if possible, at each trick (otherwise, a player may play any card); (b) a player who wins a trick must lead at the next trick. Nine tricks had already been played and there were four more tricks left to be played.

1. The distribution of suits in the four hands was as follows:
   Ada's hand: club – heart – diamond – spade
   Bea's hand: club – heart – heart – diamond
   Cyd's hand: club – heart – diamond – diamond
   Deb's hand: club – spade – spade – spade
2. Everyone followed suit when one player led a club.
3. Only one player followed suit at each of two tricks.
4. A diamond was led at trick number ten.
5. A different person led at each of the last four tricks.
6. A different person won at each of the last four tricks.
7. A different suit was led at each of the last four tricks.
8. The highest card of the suit led won each trick.

Which one of the four women led the club?

HINT: Determine the suit played by each player at each trick. The suits were led in what order?

43

Twelve C's

In the multiplication problem below, each letter represents a different digit:

$$
\begin{array}{ccccccccc}
 & A & B & C & D & E & F & G & H \\
 & \times & & & & & & A & J \\
\hline
E & J & A & H & F & D & G & K & C \\
B & D & F & H & A & J & E & C & \\
\hline
C & C & C & C & C & C & C & C & C
\end{array}
$$

Which one of the ten digits does C represent?

HINT: *A* and *B* represent which digits? Which letters represent 0 and 1?

44

John's Ideal Woman

John's ideal woman is blonde, blue-eyed, slender, and tall. He knows four women: Adele, Betty, Carol, and Doris. Only one of the four women has all four characteristics that John requires.

1. Only three of the women are both blue-eyed and slender.
2. Only two of the women are both blonde and tall.
3. Only two of the women are both slender and tall.
4. Only one of the women is both blue-eyed and blonde.
5. Adele and Betty have the same color eyes.
6. Betty and Carol have the same color hair.
7. Carol and Doris have different builds.
8. Doris and Adele are the same height.

Which one of the four women satisfies all of John's requirements?

HINT: It is not immediately evident how many of the women have each characteristic, nor how the characteristics are distributed among the four women. Each of the unidentified women referred to in some of the statements may be referred to more than once. Note the general fact that two women have different builds.

45

The L-shaped Table

Abel and his wife, Babe, gave a party to which they invited four married couples. The four men invited were: Cain, Ezra, Gene, and Ivan. The four women invited were: Dido, Fifi, Hera, and Joan.

While they were all seated at the table, one person stood up, produced a gun, and killed another person. The chairs were arranged around the oddly shaped table as in the diagram below:

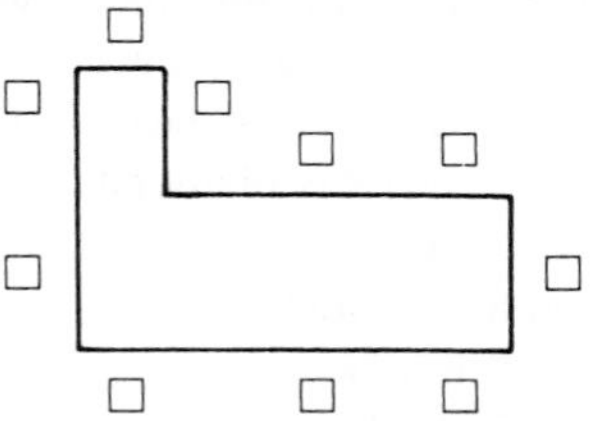

1. The people sat in alphabetical order going around the table clockwise.
2. The killer and the victim sat across the table from one another in the two newest chairs.
3. The killer's spouse and the victim's spouse sat across the table from one another in the two oldest chairs.
4. The only married couple to sit next to each other were the host and hostess.
5. The victim did not sit next to the killer's spouse.
6. The host sat alone at one side of the table.
7. The killer did not sit alone at one side of the table.
8. Both the killer and the victim were guests.

Which one of the ten was the killer?

HINT: From the three possible positions of the host, from the sets of married couples possible, and from the relative positions of the killer and victim and their spouses, the correct seating arrangement can be found. Then the killer can be determined by his or her position.

46

The Tenth Trick

Four men were playing a card game in which (a) a player must play a card in the suit led, if possible, at each trick (otherwise, a player may play any card); (b) a player who wins a trick must lead at the next trick. Nine tricks had already been played and there were four more tricks left to be played.

1. The distribution of suits in the four hands was as follows:
   HAND I: club – diamond – spade – spade
   HAND II: club – diamond – heart – heart
   HAND III: club – heart – diamond – diamond
   HAND IV: club – heart – spade – spade
2. Art led a diamond at one trick.
3. Bob led a heart at one trick.
4. Cab led a club at one trick.
5. Dan led a spade at one trick.
6. A "trump" won each trick. (A trump is any card in a certain suit that may be (a) played when a player has no cards in the suit led—in this event a card in the trump suit beats all cards in the other three suits; or (b) led, as any other suit may be led.)
7. Art and Cab, who were partners, won two tricks; Bob and Dan, who were partners, won two tricks.

Which one of the four men won the tenth trick?

HINT: Disregarding the names of the players, how many trump cards did each player have and how many tricks did each player win? Considering the leads made, what hands could each player have held? Which suit was trump?

47

Anthony's Position

Anthony, Bernard, and Charles entered some track and field events.

1. In each event three, two, and one points were awarded for first second, and third positions, respectively.
2. Duplicate points were given to those men who tied in any of these three positions.
3a. The total number of points scored by each of the three men in the pole vault, broad jump, and high jump events was the same as that scored by each of the other two men.
3b. This number of points was the same as the total number of points scored by the three men in each of these events.
4. There were no ties in the pole vault event.
5. Anthony and Charles tied in the broad jump event.
6. Anthony and Bernard tied in the high jump event.
7. Bernard scored no points in one of the three events and Charles scored no points in one of the three events.

In which one of the three positions—first, second, or third—did Anthony finish in the pole vault event?

HINT: Finding a three-by-three square array in which the totals of the columns and rows equal the same number determines Anthony's position in the pole vault event. Represent algebraically the number of points scored by Anthony and Charles in the broad jump, and by Anthony and Bernard in the high jump.

48

The Baseball Pennant

It was the last week of the baseball season and in the Felidæ League the Alleycats, the Bobcats, the Cougars, and the Domestics were all tied for first place. It was decided that there would be a series of "playoff" games; the team winning the most playoff games would win the pennant.

1. The distribution of runs scored by each team during the playoffs was as follows (listed alphabetically according to the teams' home cities):

	RUNS SCORED IN EACH OF THREE GAMES
Sexton-City team:	1 – 3 – 7
Treble-City team:	1 – 4 – 6
Ulster-City team:	2 – 3 – 6
Verdue-City team:	2 – 4 – 5

2. Each team won a different number of playoff games.
3. The score for each playoff game was different from that of any other playoff game.
4. The greatest difference in runs scored by two teams at any one game was 3 runs; this difference occurred only once when the team that lost the greatest number of playoff games lost by 3 runs.
5. Two teams scored the same number of runs during the first round and two teams scored the same number of runs during the second round of the playoffs. (A "round" consists of all the teams playing games simultaneously.)
6. During the last round the Alleycats scored the larger odd number of runs, the Bobcats scored the smaller odd number of runs, the Cougars scored the larger even number of runs, and the Domestics scored the smaller even number of runs.

Which one of the four teams—the Alleycats, the Bobcats, the Cougars, or the Domestics—won the pennant?

HINT: How many games were played altogether? Disregarding the names of the teams, how many games did each team win? Which three-game combinations of runs are possible for the pennant-winning team? Which three-game combinations of runs are possible for the team that lost the most games? What scores are possible in the games played by the pennant-winning team? What scores are possible in the games played by the team that lost the most games?

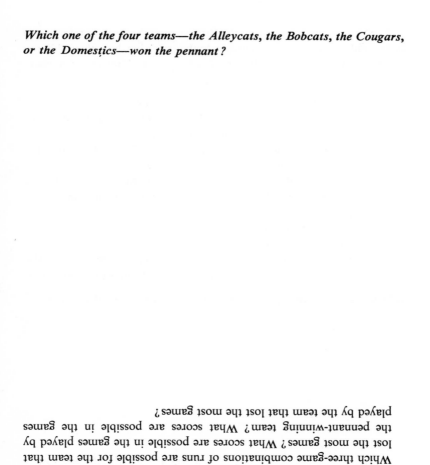

49

No Cause for Celebration

The Smiths, the Joneses, and the Browns have the habit of referring to their children by number, according to the order in which their children were born. The following statements refer to their children in this manner:

1. The second child born to each family has three brothers.
2. The third child born to each family has two sisters.
3. The fourth child born to the Joneses and the fourth child born to the Smiths have the same number of brothers.
4. The fifth child born to the Smiths and the fifth child born to the Browns have the same number of sisters.
5. The sixth child born to the Browns has the same number of brothers as the sixth child born to the Joneses has sisters.
6. Each family has a different number of children.
7. In only one family is the first-born a boy.
8. In only one family is the last-born a girl.
9. On the day that a first-born boy married a last-born girl, only the bride's family and the groom's family had cause for celebration.

Which one of the three families had no cause for celebration that day?

HINT: What are the possible boy-girl combinations for the second and third children of each family? What are the possible numbers of boys and of girls in each family? What are the possible ordered distributions of boys and girls in each family?

50

The One-Dollar Bill

Shortly after opening, the small diner contained only three customers, all men, and the owner, a woman. When the three men simultaneously got up to pay their checks, the following facts were discovered:

1. Each of the four had at least one coin, none of which was a penny or a silver dollar.
2. None of the four could change any coin.
3. The man named Lou had the largest check to pay, the man named Moe had the next-to-largest check to pay, and the man named Ned had the smallest check to pay.
4. None of the three men could give the woman any coins to pay his check and receive the correct change.
5. If the three men made equal exchanges among themselves, then each man could pay his check without requiring change.
6. When the three men made these equal exchanges they discovered that each of them then had only denominations of coins he had not held originally.

Further developments revealed these additional facts:

7. After the checks were paid and two of the men had gone, the remaining man bought some candy; this man would have been able to pay for the candy from the coins he had left, but the woman could not give him the correct change from the coins she now had.
8. The remaining man was able to pay for the candy with a one-dollar bill, but had to receive as change all the coins the woman had.

Disregarding the ensuing problems that the woman had in making change that day, ***which one of the three men gave the woman the one-dollar bill?***

56

Hint: One man had a value in coins expressible in two sets of completely different denominations, where neither set contained change for any coin. A second man made two exchanges, one with each of the other two men. This second man must be able to exchange those original coins he has left after the first exchange for coins whose denominations he did not have originally. What coin denominations could the woman have held before and after the checks were paid? The amounts of the three checks, the cost of the candy, and the value of the woman's coins before the checks were paid totaled how much?

SOLUTIONS

1 / The Best Player

The best player and the best player's twin are the same age; the best player and the worst player are the same age, from [2]; and the best player's twin and the worst player are two different people, from [1]. Therefore, three of the four people are the same age. So Mr. Scott's son, daughter, and sister must be the same age, since Mr. Scott must be older than both his son and his daughter. Then Mr. Scott's son and daughter must be the twins indicated in [1].

So either Mr. Scott's son or daughter is the best player and Mr. Scott's sister is the worst player. Since the best player's twin must be Mr. Scott's son, from [1], ***the best player must be Mr. Scott's daughter.***

2 / Middletown

The man who traveled along all the streets in Middletown exactly once (a) had to pass through his home intersection an odd number of times (from [3]), in order to leave it last; and (b) had to pass through his friend's home intersection an odd number of times (from [3]), in order to enter it last. So this man's home was located at an intersection with an odd number of streets and his friend's home was located at an intersection with an odd number of streets.

Then, from [1], either Arden visited Duane or Duane visited Arden. From [2], since Arden did not visit Duane and Duane did visit Arden, ***Duane must be the man who traveled along all the streets in Middletown exactly once.***

A possible layout of Middletown is shown on page 58. The layout is drawn in one continuous line to indicate a possible route of Duane.

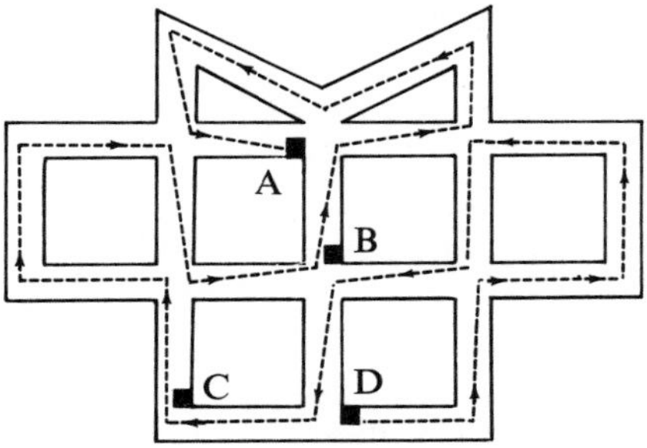

3 / Murder in the Family

The youngest member was not the victim, from [3]; was not the accessory, from [4]; and was not the killer, from [6]. So, from [4], there are only three possibilities (*A* stands for accessory, *V* for victim, *K* for killer, and *W* for witness):

	I	II	III
oldest member	*A*	*A*	*K*
next-to-oldest member	*V*	*K*	*A*
next-to-youngest member	*K*	*V*	*V*
youngest member	*W*	*W*	*W*

From [5], the father was the oldest member; so the next-to-oldest member was the mother. From [2] and the above possibilities, the youngest member was the daughter; so the next-to-youngest member was the son. Then from oldest member to youngest member the three possibilities are:

	I	II	III
father	*A*	*A*	*K*
mother	*V*	*K*	*A*
son	*K*	*V*	*V*
daughter	*W*	*W*	*W*

From [3], I is impossible. From [1], III is impossible. So II becomes the only remaining possibility. Therefore, ***the mother was the killer.***

4 / Three A's

A cannot equal 0 because then M and N would equal 0.

A cannot equal 1 because the product is different from AS.

A cannot equal 2 because a three-digit product would not be possible.

A cannot equal 3 because 4 cannot be carried to $A \times A$.

A cannot equal 4 or 7 because 8 cannot be carried to $A \times A$.

A cannot equal 5 or 6 because then S would have to equal 0, making N equal to S, or S would have to equal 1, making N equal to A.

A cannot equal 9 because then 8 would have to be carried, making A equal to S.

So ***A must be equal to 8.***

Though not necessary for the solution of the problem, the numerical values of S, M, and N can now be determined: since 4 must be carried, S equals 5 or 6; but S cannot equal 6 because then A would equal N. So S is equal to 5. The multiplication is shown below.

$$\begin{array}{r} 8\ 5 \\ \times\ 8 \\ \hline 6\ 8\ 0 \end{array}$$

5 / Mary's Ideal Man

From [1], three of the men are tall and one is not. Then, from [4], Bill and Carl are both tall. Then, from [5], Dave is not tall.

From [2], Dave must have at least one of the required traits; since he is not tall, he must be dark. (Only Mary's ideal man is handsome, but her ideal man is also dark.)

From [1], only two of the men are dark. Then, from [3], Alec and Bill are either both dark or both not dark. Since Dave is dark, Alec and Bill are not dark, otherwise three men would be dark. From [1], and the fact that Dave is dark, Carl must be dark.

Since Dave is not tall, Alec and Bill are not dark, and Carl is both tall and dark, ***Carl is the only man who could be Mary's ideal man*** (so he must be handsome).

IN SUMMARY: Alec is tall.
Bill is tall.
Carl is tall, dark, and handsome.
Dave is dark.

6 / Everybody Lied

From [II], negating each of the eight false statements results in the following eight true statements:

[1] One of the four killed the psychiatrist.
[2] The psychiatrist was dead when Avery left.
[3] Blake was not the second to arrive.
[4] The psychiatrist was alive when Blake arrived.
[5] Crown was not the third to arrive.
[6] The psychiatrist was dead when Crown left.
[7] The killer arrived after Davis did.
[8] The psychiatrist was alive when Davis arrived.

From [I], [4], [8], [2], and [6], Blake and Davis arrived before Avery and Crown. From [3], Davis must have arrived second; so Blake arrived first. From [5], Avery must have arrived third; so Crown arrived fourth.

The psychiatrist was alive when Davis arrived second but was dead when Avery left third. So, from [1], either Avery or Davis killed the psychiatrist.

From [7], ***Avery is the murderer.***

7 / The Triangular Pen

From [1], [2], [3], and [6], the lengths of the sides of the triangular pen are in the ratio of 1 to 2 to 3, one number of which is incorrect.

From [4], the incorrect number can be replaced only by a whole number.

From [5], the incorrect number must be replaced by a whole number greater than 3. If either 2 or 3 were replaced by a whole number greater than 3, it would be impossible to construct the pen, since the sum of the lengths of any two sides must be greater than the length of the third side. So 1 is the incorrect number and ***the \$10 cost for the side of the pen facing the barn must be incorrect.***

If 1 is replaced by a whole number greater than 4 the pen would still be impossible to construct. However, if 1 is replaced by 4, the pen can be constructed. The cost for the side facing barn must have been \$40 instead of \$10.

8 / Speaking of Bets

Let a be the amount A had and b be the amount B had before A and B bet. Then, from [1], after they bet A had $2a$ and B had $b - a$.

Let c be the amount C had before he bet with B. Then, from [2], after B and C bet, B had $(b - a) + (b - a)$ or $2b - 2a$, and C had $c - (b - a)$ or $c - b + a$.

Then, from [3], after C and A bet, C had $(c - b + a) + (c - b + a)$ or $2c - 2b + 2a$, and A had $2a - (c - b + a)$ or $a - c + b$.

From [4], $a - c + b = 2b - 2a$ and $a - c + b = 2c - 2b + 2a$. The first equation yields: $b = 3a - c$, and the second equation yields: $3b = a + 3c$. Multiplying the first of these latter equations by 3 and adding the two equations yields: $6b = 10a$ or $b = \frac{5}{3}a$. Substitution for b yields: $c = \frac{4}{3}a$.

So A started with a cents, B with $\frac{5}{3}a$ cents, and C with $\frac{4}{3}a$ cents.

From [5], a cannot be 50 cents because then B and C would have started with fractions of a cent, and $\frac{4}{3}a$ cannot be 50 cents because then A and B would have started with fractions of a cent. So $\frac{5}{3}a$ is 50 cents and ***B is the speaker.***

In summary, A started with 30 cents, B started with 50 cents, and C started with 40 cents.

9 / The Trump Suit

From [1], [2], and [3], the distribution of the four suits is either:

(a) 1 2 3 7

or

(b) 1 2 4 6

or

(c) 1 3 4 5

From [6], combination (c) is eliminated because no suit consists of only two cards.

From [5], combination (a) is eliminated because the addition of no two numbers produces a sum of six.

So (b) is the correct combination of suits.

From [5], either there are two hearts and four spades or there are four hearts and two spades.

From [4], either there are one heart and four diamonds or there are four hearts and one diamond.

From [4] and [5] together, there must be four hearts. Then there must be two spades. So ***spades is the trump suit.***

In summary, there are four hearts, two spades, one diamond, and six clubs.

10 / Malice and Alice

From [1], [2], and [3], the roles of the five people were as follows:

Man in bar	Killer on beach	Child alone
Woman in bar	Victim on beach	

Then, from [4], either Alice's husband was in the bar and Alice was on the beach, or Alice was in the bar and Alice's husband was on the beach.

If Alice's husband was in the bar, the woman he was with was his daughter, the child who was alone was his son, and Alice and her brother were on the beach. Then either Alice or her brother was the victim; so the other was the killer. But, from [5], the victim had a twin and this twin was innocent. Since Alice and her brother could only be twins to each other, this situation is impossible. Therefore Alice's husband was not in the bar.

So Alice was in the bar. If Alice was in the bar, she was with her brother or her son.

If she was with her brother, her husband was on the beach with one of the two children. From [5], the victim could not be her husband, because none of the others could be his twin; so the killer was her husband and the victim was the child. But this situation is impossible, because it contradicts [6]. Therefore Alice was not with her brother in the bar. So Alice was with her son in the bar. Then the child who was alone was her daughter. Therefore, Alice's husband was with Alice's brother on the beach. From previous reasoning, the victim could not be Alice's husband. But the victim could be Alice's brother, because Alice could be his twin. So ***Alice's brother was the victim.***

11 / A Week in Arlington

If Sunday is the first of the six consecutive days mentioned, then from [1], [2], and [4], the supermarket is closed only on Sunday, Monday, and Wednesday. This is impossible, from [3].

If Monday is the first of the six consecutive days mentioned, then from [2] and [4] at least one place is closed each day. This is impossible because all three places are open together on one day each week.

If Tuesday is the first of the six consecutive days mentioned, then from [1], [2], and [4] the department store is closed only on Tuesday, Saturday, and Sunday. This is impossible, from [3].

If Wednesday is the first of the six consecutive days mentioned, then from [1], [2], and [4] the bank is closed only on Sunday, Monday, and Friday, and the supermarket is closed only on Sunday, Thursday, and Saturday. This is impossible, from [3].

If Thursday is the first of the six consecutive days mentioned, then from [1], [2], and [4] the bank is closed only on Tuesday, Saturday, and Sunday. This is impossible, from [3].

If Friday is the first of the six consecutive days mentioned, then from [1], [2], and [4] the supermarket is closed only on Monday, Saturday, and Sunday. This is impossible, from [3].

So Saturday is the first of the six consecutive days mentioned; then from [1], [2], and [4] (*C* stands for closed, *O* for open):

	SUN	MON	TUES	WED	THURS	FRI	SAT
Bank	*C*	*C*	*O*	*O*	*C*	*O*	*O*
Department store	*C*	*O*	*O*	*C*	*O*	*O*	*C*
Supermarket	*C*	*O*	*C*				

From the above table, ***Friday must be the day all three places are open.***

To complete the table: from [1] and [3], the supermarket cannot be closed on Wednesday or Saturday; so the supermarket must be closed on Thursday.

12 / Eunice's Marital Status

From [1] and [2], six couples came to the party. From [3], [4], and [5], if a equals the number of married women present, then $6 - a$ equals the number of engaged women present and $6 - a$ equals the number of engaged men present.

From [6], then, $6 - a$ equals the number of married men present.

If b equals the number of married men who came alone, then the number of married men who came with their wives (a) plus the number of married men who came alone (b) equals the total number of married men present: $a + b = 6 - a$. Then the number of married men who came alone (b) equals $6 - 2a$.

From [7], $6 - 2a$ equals the number of unattached men who came alone.

From [4], the number of unattached women who came alone, then, equals the number of people who came alone (7) minus the number of married men who came alone ($6 - 2a$) minus the number of unattached men who came alone: $7 - (6 - 2a) - (6 - 2a)$ or $4a - 5$.

So a equals the number of married women present, $6 - a$ equals the number of engaged women present, and $4a - 5$ equals the number of unattached women present.

Since $4a - 5$ equals the number of unattached women present, a cannot equal zero or one. From [9], Jack was unattached; so a cannot be greater than 2, otherwise the number of unattached men ($6 - 2a$) would be 0 or less. Therefore, a must equal 2.

So there were two married women, four engaged women, and three unattached women at the party.

From [8], ***Eunice was an engaged woman.***

13 / A Smart Man

If Aaron is smart, then from [2] he will pass Chemistry and from [II] he will not pass Physics. If Aaron is not smart, then from [1] he will not pass Physics and from [II] he will pass Chemistry.

If Brian is smart, then from [4] he will pass Physics and from [II] he will not pass Chemistry. If Brian is not smart, then from [3] he will not pass Chemistry and from [II] he will pass Physics.

If Colin is smart, then from [6] he will pass Physics and from [II] he will not pass Chemistry. If Colin is not smart, then from [5] he will not pass Physics and from [II] he will pass Chemistry.

The following is now known:

IF	THEN HE PASSES ONLY
Aaron is smart	Chemistry
Aaron is not smart	Chemistry
Brian is smart	Physics
Brian is not smart	Physics
Colin is smart	Physics
Colin is not smart	Chemistry

Aaron cannot be the only smart man because then Aaron and Colin will pass Chemistry, contradicting [I]. Colin cannot be the only smart man because then Brian and Colin will pass Physics, contradicting [I]. However, if Brian is the only smart man, then he is the only man to pass Physics, satisfying [I], and he is the only man not to pass Chemistry, satisfying [II].

So ***Brian is smart.***

14 / The Murderer

From [I], each statement was made by a suspect not mentioned in the statement. Therefore, there are only two ways the statements could have been made:

A		B	
1. Brad:	Adam is innocent.	1. Cole:	Adam is innocent.
2. Cole:	Brad is telling the truth.	2. Adam:	Brad is telling the truth.
3. Adam:	Cole is lying.	3. Brad:	Cole is lying.

For *A*: [2] supports [1]; and [3], denying [2], denies [1]. In effect, the statements become:

1. Brad: Adam is innocent.
2. Cole: Adam is innocent.
3. Adam: Adam is guilty.

If "Adam is guilty" is true, then Adam told the truth and is guilty. This is impossible, from [II].

If "Adam is innocent" is true, then Brad and Cole told the truth, and one of them is guilty. This is impossible, from [II].

So A must be incorrect.

For B: [3] denies [1]; and [2], supporting [3], denies [1]. In effect, the statements become:

1. Cole: Adam is innocent.
2. Adam: Adam is guilty.
3. Brad: Adam is guilty.

If "Adam is guilty" is true, then Adam told the truth and is guilty. This is impossible, from [II].

If "Adam is innocent" is true, then Adam is innocent, and Adam and Brad lied. Since only Adam and Brad lied, one of them is guilty. Since Adam is innocent (even though he lied), ***Brad must be the murderer.***

15 / The Missing Digit

Since there are four different digits to be added in each column, the largest sum for a column is $9 + 8 + 7 + 6$ or 30. Since I cannot equal zero, no more than 2 can be carried to the left column. Since no more than 2 can be carried to the left column, I cannot equal 3. So I must equal 1 or 2.

If I equals 1, then the right column must add to 11 or 21, while the left column must add to 10 or 9, respectively. Then

$$(B + D + F + H) + (A + C + E + G) + I = 11 + 10 + 1 = 22$$

or

$$(B + D + F + H) + (A + C + E + G) + I = 21 + 9 + 1 = 31$$

But since the sum of the digits 0 through 9 is 45, this situation is impossible, because the difference between the sum of the ten digits and the nine digits present in the addition is greater than 9 ($45 - 22$ or $45 - 31$). So I must be equal to 2.

Since I equals 2, the right column must add to 12 or 22 while the left column must add to 21 or 20, respectively. Then

$$(B + D + F + H) + (A + C + E + G) + I = 12 + 21 + 2 = 35$$

or

$$(B + D + E + H) + (A + C + E + G) + I = 22 + 20 + 2 = 44.$$

The first alternative is impossible, because the difference between the sum of the ten digits and the nine digits present in the addition is greater than 9 (45 − 35). So ***the missing digit must be 1*** (45 − 44).

The existence of at least one addition combination can be confirmed as follows: since zero cannot occur in the left column, by convention, it must occur in the right column. Then three digits must total 22 in the right column. There are only two possibilities for the four digits: either 0, 5, 8, and 9 or 0, 6, 7, and 9. The four digits in the left column, then, are either: 3, 4, 6, and 7 (with 0, 5, 8, and 9 in the right column) or 3, 4, 5, and 8 (with 0, 6, 7, and 9 in the right column). Thus, many addition combinations are possible.

16 / One, Two, or Four

From [II], if a player *can* win, he *must* win.
From [2] and [3]:

(a) Drawing from one coin, a player loses.
(b) Drawing from two coins, a player wins by taking only one coin, thus putting the other player in the losing position of drawing from one coin.
(c) Drawing from three coins, a player wins by taking two coins, putting the other player in the same losing position as in (b). If he takes only one coin, the other player may take only one coin and win.
(d) Drawing from four coins, a player loses. If he takes one coin he gives the other player the winning position of drawing from three coins. If he takes two coins he gives the other player the winning position of drawing from two coins. If he takes four coins he loses at once. He cannot win because he cannot leave a number of coins that represents a losing position for the other player.

(e) Drawing from five coins, a player wins if he is able to leave a number of coins that represents a losing position for the other player. So if he can leave one coin or four coins for the other player to draw from he wins. Accordingly, he takes four coins, leaving one, or one coin, leaving four.

Reasoning in this manner, one finds that drawings from one, four, seven, and ten coins are losing positions, and drawings from two, three, five, six, eight, and nine coins are winning positions. The following tables summarize how these two sets of drawings can be losing and winning positions, respectively:

FROM A LOSING POSITION OF	IF A PLAYER DRAWS	HE LEAVES A WINNING POSITION OF
4	1	3
	2	2
	4	0
7	1	6
	2	5
	4	3
10	1	9
	2	8
	4	6

FROM A WINNING POSITION OF	DRAW (1, 2, or 4)	TO LEAVE A LOSING POSITION OF
2	1	1
3	2	1
5	1	4
	4	1
6	2	4
8	1	7
	4	4
9	2	7

From [1], there are ten coins. Since drawing from ten coins is a losing position, whoever goes first must lose. Since Austin goes first, from [I], Austin must lose. So ***Brooks must win.***

17 / The Student Thief

From [6] and [4], Cobb attended two classes not conducted by the professor. From [6] and [3], Burt attended one class not conducted by the professor. From [6] and [2], Amos attended only classes conducted by the professor. If *P* represents a class conducted by the professor and *O* represents a class not the professor's, then, from [1] and [5], the following table can be constructed (an *x* indicates attendance at a class):

	AMOS	BURT	COBB
P			
P			
P			
O		*x*	*x*
O			*x*

From [6] and [7]—applying [7] only to the professor's classes for now—four records of attendance are possible, as shown below.

	I			II		
	AMOS	BURT	COBB	AMOS	BURT	COBB
P	*x*	*x*		*x*		
P		*x*	*x*		*x*	*x*
P	*x*		*x*	*x*	*x*	*x*
O		*x*	*x*		*x*	*x*
O			*x*			*x*

	III			IV		
	AMOS	BURT	COBB	AMOS	BURT	COBB
P		*x*				*x*
P	*x*		*x*	*x*	*x*	
P	*x*	*x*	*x*	*x*	*x*	*x*
O		*x*	*x*		*x*	*x*
O			*x*			*x*

Then—applying [7] to all five classes—possibilities I, II, and IV are eliminated.

From III and from [8], the two students who were innocent must be Amos and Cobb (only one of the professor's three classes was attended by two students).

Therefore, ***Burt stole the answer key.***

18 / The Four Groves

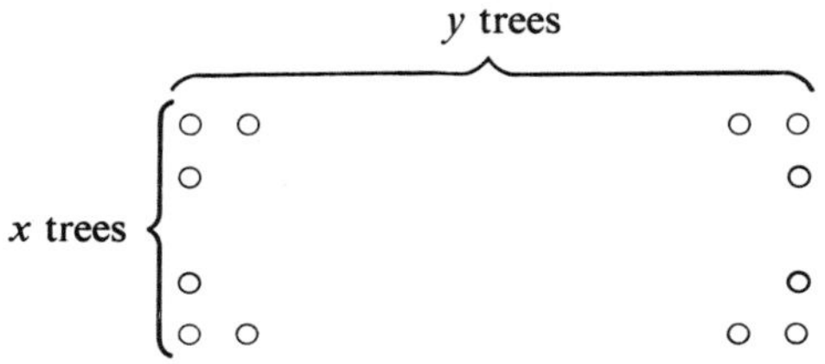

From [1], let x = the number of trees along each of two opposite sides of the three groves mentioned in [3], and let y = the number of trees along each of the other two opposite sides. Then the number of trees on the border is equal to $y + y + (x - 2) + (x - 2)$ or $2y + 2x - 4$, and the number of trees in the interior is equal to $(x - 2)$ multiplied by $(y - 2)$.

From [3], $2y + 2x - 4 = (x - 2)(y - 2)$. Solving for x, $x = (4y - 8)/(y - 4)$. Thus y must be greater than 4 and $(y - 4)$ must divide exactly into $4y - 8$. Trial and error quickly yields four pairs of values.

x	y
12	5
8	6
6	8
5	12

(These are the only possible values because $(4y - 8)/(y - 4)$ is equal to $4 + 8/(y - 4)$; in order that $8/(y - 4)$ be a positive whole number, y can only be 5, 6, 8, or 12.)

From [2], the apple grove must contain 5 rows, the lemon grove must contain 6 rows, the orange grove must contain 7 rows, and the peach grove must contain 8 rows.

Since it is not possible for a grove with 7 rows to satisfy condition [3], ***the number of trees in the interior of the orange grove does not equal the number of trees on its border.***

In summary, the apple grove has 5 rows of 12 trees each (30 trees on the border and 30 trees in the interior), the lemon grove has 6 rows of 8 trees each (24 trees on the border and 24 trees in the interior), and the peach grove has 8 rows of 6 trees each (the same size as the lemon grove).

19 / Hubert's First Watch

From [1], no more than 100 days occurred between Hubert's first and last watches.

From [2], the number of days occurring between Hubert's first and last watches must be a multiple of seven.

From [3] and [4], February is not the month Hubert had his first watch because no other month can have the same number of days as February. So more than 28 days occurred between Hubert's first and last watches.

From the previous statements, the number of days occurring between Hubert's first and last watches must be one of the following: 35, 42, 49, 56, 63, 70, 77, 84, 91, or 98.

From these ten possibilities, more than one month and less than four months passed between Hubert's first and last watches. So either two or three months passed between Hubert's first and last watches.

MONTH	NUMBER OF DAYS IN THE MONTH	NUMBER OF DAYS IN TWO CONSECUTIVE MONTHS BEGINNING WITH THE MONTH ON THE LEFT	NUMBER OF DAYS IN THREE CONSECUTIVE MONTHS BEGINNING WITH THE MONTH ON THE LEFT
January	31	59 *or* 60	90 *or* 91
February	28 *or* 29	59 *or* 60	89 *or* 90
March	31	61	92
April	30	61	91
May	31	61	92
June	30	61	92
July	31	62	92
August	31	61	92
September	30	61	91
October	31	61	92
November	30	61	92
December	31	62	90 *or* 91

In the above table, the only one of the ten possibilities that occurs is 91 days. So 91 days passed between Hubert's first and last watches, resulting in four possible pairs of months for the first and last watches.

	MONTH OF FIRST WATCH	MONTH OF LAST WATCH
I	January (31 days)	April (30 days)
II	April (30 days)	July (31 days)
III	September (30 days)	December (31 days)
IV	December (31 days)	March (31 days)

From [4], ***December must have been the month of Hubert's first watch.***

20 / The Square Table

PART I

If [6] is false, then Clark did not poison Doyle and statements [1] through [4] are true, from [7]. Then the seating arrangement (using the first letter of each name) was

A
B D
C

and, from [2] and [4] respectively, Brent and Alden are innocent. From [8], this situation is impossible.

So [6] is true. Statements [1] and [5] cannot both be true; so from [6] and [7] either Alden or Clark is guilty, and [3] and [4] are true. From [3] and [4], the seating arrangement was one of the following:

I	II	III
C B A D	C B D A	A B D C

In the first arrangement, since [4] is true, Alden is innocent; so from [6] and [7] Alden told the truth. But [1] is not true for this arrangement, so this arrangement is impossible.

In the second arrangement, since [4] is true, Clark is innocent; so from [6] and [7] Clark told the truth. But [5] is not true for this arrangement, so this arrangement is impossible.

The third arrangement must be the correct one, because it is the only one possible. Since [4] is true, Alden is innocent. Then ***Clark must have poisoned Doyle.***

From [6] and [7], Alden told the truth. So the truth of [1] is consistent with the seating arrangement and the truth of [2] is consistent with Brent's established innocence. Then [5] must be false, from [7]. That [5] is false is consistent with the seating arrangement.

The Square Table

PART II

If [3] is false, then the seating arrangement (using the first letter of each name) was either:

	I				II	
	R				D	
S		T	*or*	S		T
	D				R	

If [3] is false, then [6] is true, from [7]. So Sid is the murderer. From [7], statements [1], [2], and [5] are also true.

In the first arrangement [2] is false (Sid is the murderer), so this arrangement is impossible.

In the second arrangement, it happens that [1], [2], and [5] are true. However, from the statements in Part I the murderer sat on Doyle's left; in this arrangement, Sid sits on Doyle's right. So this arrangement is impossible.

So [3] is true.

If [1] is true, then [5] is true. If [5] is true, then [1] is true. From [7], [1] and [5] cannot both be false. So [1] and [5] are both true.

Then [1], [3], and [5] are all true.

Since [1], [3], and [5] are all true, the seating arrangement was either:

III		IV
R		R
S D	*or*	D S
T		T

If [6] is true, Ray or Sid is the murderer, from [7] and the fact that [5] is true. If [6] is false, Ray or Sid is again the murderer. Since the murderer sat on Doyle's left, Ted or Ray is the murderer, from the third and fourth arrangements. Then the murderer must be Ray.

So ***Ray's wife must be the murderer's wife.***

It follows that arrangement IV is the correct arrangement. A test of the truth of each statement with this arrangement reveals that only [2] is false.

(The seating arrangement of Part I can be made to correspond to the seating arrangement of Part II by moving each man two places in the same direction around the table. Thus,

A
B D
C

in Part I corresponds to

T
S D
R

in Part II.)

21 / The Two Brothers

Let x be the price in cents of each art object bought by the two brothers and B be the total number of objects they bought; then they spent Bx cents for their art objects, by [3].

Let $2x$ be the price in cents of each art object bought by the other four men and M be the total number of art objects they bought; then they spent $M2x$ or $2Mx$ cents for their art objects, by [4].

Then the total money spent can be represented by the equation $Bx + 2Mx = 100{,}000$ or $(B + 2M)x = 100{,}000$, by [5]. Since 100,000 is wholly divisible only by multiples of 2 or 5 (excluding multiples of 1), $B + 2M$ must be a multiple of either 2 only, 5 only, or both 2 and 5 only, because x cannot contain a fraction of a cent, by [1].

The total number of art objects bought was $1 + 2 + 3 + 4 + 5 + 6 = 21$. The possible numbers of art objects bought by the two brothers (B) and the corresponding possible numbers of art objects bought by the other four men (M) are listed below. Also listed are the possible values for $2M$ and $2M + B$. (It might be noted that if the same value for B can be gotten by adding more than one pair of numbers—those combinations not checked—there will be more than one possible solution.)

B		M	$2M$	$2M + B$
√ 1 + 2	= 3	18	36	39
√ 1 + 3	= 4	17	34	38
1 + 4 2 + 3	= 5	16	32	37
1 + 5 2 + 4	= 6	15	30	36
1 + 6 2 + 5 3 + 4	= 7	14	28	35
2 + 6 3 + 5	= 8	13	26	34
3 + 6 4 + 5	= 9	12	24	33
√ 4 + 6	= 10	11	22	32
√ 5 + 6	= 11	10	20	31

Of the nine possible values for $2M + B$, just the value 32 is a multiple of 2 only, 5 only, or both 2 and 5 only.

The corresponding value for B is 10, and 10 is the sum of only one pair of numbers: 4 and 6. Therefore, from [2], ***Dwight and Farley are brothers.***

To find the cost of the art objects, substitute 32 for $2M + B$:

$$(B + 2M)x = 100{,}000$$
$$32x = 100{,}000$$
$$x = 3125$$
$$2x = 6250$$

Checking the solution:

$$(4 + 6 = 10) \quad 10 \times 3125 = 31{,}250$$
$$(1 + 2 + 3 + 5 = 11) \quad 11 \times 6250 = 68{,}750$$
$$100{,}000$$

22 / One, Three, or Four

From [II], if a player *can* win, he *must* win.
From [2] and [3]:

(a) Drawing from one coin, a player wins.
(b) Drawing from two coins, a player loses. He must take one coin, thus putting the other player in the winning position of drawing from one coin.
(c) Drawing from three coins, a player wins because if he takes all three coins he wins at once and if he takes one coin he puts the other player in the losing position of drawing from two coins.
(d) Drawing from four coins, a player wins by taking all four coins. If he takes one coin he gives the other player the winning position of drawing from three coins. If he takes three coins he gives the other player the winning position of drawing from one coin.
(e) Drawing from five coins, a player wins if he is able to leave a number of coins that represents a losing position for the other player. So if he can leave two coins for the other player to draw from he wins. Accordingly, he takes three coins.

Reasoning in this manner, one finds that drawings from two, seven, and nine coins are losing positions, and drawings from one, three, four, five, six, and eight coins are winning positions. The following tables summarize how these two sets of drawings can be losing and winning positions, respectively:

FROM A LOSING POSITION OF	IF A PLAYER DRAWS	HE LEAVES A WINNING POSITION OF
2	1	1
7	1	6
	3	4
	4	3
9	1	8
	3	6
	4	5

FROM A WINNING POSITION OF	DRAW (1, 3, *or* 4)	TO LEAVE A LOSING POSITION OF
1	1	0
3	{1	{2
	3	0
4	4	0
5	3	2
6	4	2
8	1	7

From [1], there are nine coins. Since drawing from nine coins is a losing position, whoever goes first must lose. Since Aubrey goes first, from [I], Aubrey must lose. So ***Blaine must win.***

23 / A Timely Death

From [1], Osborn was shot in his camp before 10:30. Therefore (if one of the three men shot Osborn), from [2], [4], and [6], it is not possible that Osborn was shot in his camp after 9:50 because then each of the three men would have had only 75 minutes to get to Osborn's camp and to return to his own camp: Xavier had from 9:40 to 10:55, Yeoman had from 9:45 to 11:00, and Zenger had from 9:50 to 11:05.

From [1], Osborn died instantly; from [6], Osborn replied to Wilson's call at 9:15. So (if one of the three men shot Osborn) Osborn was shot after 9:15 and no later than 9:50.

Statement [6] allows each of the three men a maximum of 85 minutes to get to Osborn's camp and to return to his own camp: Xavier had from 8:15 to 9:40, Yeoman had from 8:20 to 9:45, and Zenger had from 8:25 to 9:50. So,

If Xavier shot Osborn, he must have done so before 9:40
If Yeoman shot Osborn, he must have done so before 9:45
If Zenger shot Osborn, he must have done so before 9:50

But, from [3], [4], and [5]:

Xavier required more than 40 minutes to get to Osborn's camp
Yeoman required at least 40 minutes to get to Osborn's camp and required at least 40 minutes to return to his own camp
Zenger required more than 40 minutes to return to his own camp

If Yeoman shot Osborn, he must have left Osborn's camp by 9:05, at least 40 minutes before replying to Wilson's second call. If Zenger shot Osborn, he must have left Osborn's camp before 9:10, more than 40 minutes before replying to Wilson's second call. So if Yeoman or Zenger shot Osborn a reply from Osborn at 9:15 would not have been possible because Osborn, having died instantly, could not have answered.

However, if Xavier shot Osborn, he had only to arrive at Osborn's camp after 8:55, more than 40 minutes after replying to Wilson's first call. So if Xavier left Osborn's camp after 9:15 he still may have had enough time to return to his own camp. Therefore, ***Xavier was not eliminated from Wilson's suspicion.***

24 / Turnabout

A, *B*, and *C* occur in the same positions in the three problems. So, from [2], two problems can be considered as one problem with an indeterminate number of letters occurring between *A* and *B*. Since two products contain in reverse order the same digits as the numbers multiplied by *C*, two problems can be represented as the following single problem.

$$\begin{array}{r} A \cdots B \\ \times \quad C \\ \hline B \cdots A \end{array}$$

Then, from [1], the following reasoning can be applied to this single problem:

$C \times A$ must be less than 10, neither *C* nor *A* can be zero, and *C* cannot be 1. (Though *A* is not zero by convention, because it begins a number, it is also impossible for *A* to be zero.) So the following values for *C* and *A* are possible:

	C	*A*		*C*	*A*		*C*	*A*
(1)	2	1	(5)	3	2	(9)	6	1
(2)	2	3	(6)	4	1	(10)	7	1
(3)	2	4	(7)	4	2	(11)	8	1
(4)	3	1	(8)	5	1	(12)	9	1

$C \times B$ must end in *A*, so if *C* is even *A* must be even also. This consideration eliminates (1), (2), (6), (9), and (11). If *C* is 5, $C \times B$ can only end in 5 or 0, not 1; so (8) is eliminated. If C is 9, B must be 9 in order that A be 1; but C and B cannot both be 9, so (12) is eliminated. Multiplying the remaining possible values of C by 1 through 9 to get a value for B in order that A be the value listed, one arrives at the following combinations:

	C	B	A		C	B	A
(3)	2	7	4	(7a)	4	3	2
(4)	3	7	1	(7b)	4	8	2
(5)	3	4	2	(10)	7	3	1

$C \times A$ must be less than or equal to B; so (3), (5), (7a), and (10) are eliminated. For (4) and (7b), one arrives at these two possible partial multiplications.

$$(4)\quad \begin{array}{r} 1\ R \cdots 7 \\ \times \quad\ \ 3 \\ \hline 7\ \cdots\cdot\ 1 \end{array} \qquad (7b)\quad \begin{array}{r} 2\ R \cdots 8 \\ \times \quad\ \ 4 \\ \hline 8\ \cdots\cdot\ 2 \end{array}$$

In (4) it is impossible to carry as much as 4 from $3 \times R$ to 3×1 to get 7; so (4) is eliminated. Therefore (7b) is the only possible partial multiplication, and $A = 2$, $B = 8$, and $C = 4$ in the two problems mentioned in statement [2].

In (7b), 3 is carried from 4×8. So, if R occurs in the product, R cannot equal zero. R cannot equal 2 because $A = 2$. R cannot equal more than 2 because nothing is carried from $4 \times R$ to 4×2. So $R = 1$, if R occurs in the product.

In order for the digits in the product to be the same as the digits in the number multiplied by C, it is necessary for R to be equal to 1. If 1 is substituted for R in problem I, the multiplication becomes:

$$\begin{array}{r} 2\ 1\ 8 \\ \times\ \ \ 4 \\ \hline 8\ 7\ 2 \end{array}$$

The product 872 is not the reverse of 218; so, from [2], ***problem* I *is the problem in which the digits in the product are not the same as the digits in the number multiplied by C.***

Problems II and III can be completed as follows:

In problem II,

$$\begin{array}{r} 2\,1\,S\,8 \\ \times \quad 4 \\ \hline 8\,S\,1\,2 \end{array}, S \text{ must be } 7: \begin{array}{r} 2\,1\,7\,8 \\ \times \quad 4 \\ \hline 8\,7\,1\,2 \end{array}$$

In problem III,

$$\begin{array}{r} 2\,1\,S\,T\,8 \\ \times \quad 4 \\ \hline 8\,T\,S\,1\,2 \end{array}, T \text{ must be } 7: \begin{array}{r} 2\,1\,S\,7\,8 \\ \times \quad 4 \\ \hline 8\,7\,S\,1\,2 \end{array}$$

Then S must be 9: $$\begin{array}{r} 2\,1\,9\,7\,8 \\ \times \quad 4 \\ \hline 8\,7\,9\,1\,2 \end{array}$$

25 / A Week in Burmingham

From [3] and [4], the supermarket is open on Monday, Tuesday, Thursday, Friday, and Saturday, and is closed on Sunday and Wednesday.

From [5], on the third consecutive day the supermarket is closed. Therefore, the first consecutive day in [5] must be Friday or Monday.

From [6], on the second consecutive day the supermarket is closed. Therefore, the first consecutive day in [6] must be Tuesday or Saturday.

So for the two sequences in [5] and [6] there are the following possibilities:

	SEQUENCE IN [5] BEGINS ON	*and*	SEQUENCE IN [6] BEGINS ON
I.	Friday		Tuesday
II.	Friday		Saturday
III.	Monday		Tuesday
IV.	Monday		Saturday

From [3], [4], [5], and [6], the following four schedules result (*C* stands for closed, *O* for open):

	SUN	MON	TUES	WED	THURS	FRI	SAT	
Bank	*C*		*C*	*C*			*C*	
Department store	*C*			*C*	*C*	*C*		I
Supermarket	*C*	*O*	*O*	*C*	*O*	*O*	*O*	

	SUN	MON	TUES	WED	THURS	FRI	SAT
Bank	*C*			*C*			*C*
Department store	*C*	*C*		*C*		*C*	
Supermarket	*C*	*O*	*O*	*C*	*O*	*O*	*O*

II

	SUN	MON	TUES	WED	THURS	FRI	SAT
Bank	*C*		*C*	*C*			
Department store	*C*	*C*		*C*	*C*		
Supermarket	*C*	*O*	*O*	*C*	*O*	*O*	*O*

III

	SUN	MON	TUES	WED	THURS	FRI	SAT
Bank	*C*		*C*	*C*			*C*
Department store	*C*	*C*		*C*			
Supermarket	*C*	*O*	*O*	*C*	*O*	*O*	*O*

IV

Schedules I, II, and III contradict [2]. So IV is the correct schedule.

From [2], the department store is open the remaining days; so, from [1], the bank is closed on Thursday and Friday. Since the bank was open when I went into Burmingham, ***I must have gone into the town on Monday.***

26 / Cards on the Table

From [1] and [2], one and only one of the following statements is true:

(a) card 3 is a Queen and card 6 is a Queen
(b) only card 3 is a Queen
(c) only card 6 is a Queen
(d) only card 4 is a Queen

If card 3 is a Queen and card 6 is a Queen, then either of the following partial arrangements must occur (*K* represents a King, *Q* represents a Queen, and *X* represents an unknown card):

		X					*X*	
K	*Q*	*K*			*K*	*Q*	*K*	
	K	*Q*	*K*			*X*	*Q*	*X*
		X					*K*	

Since [3] cannot be satisfied in either arrangement, situation (a) is impossible.

If only card 3 is a Queen, then card 6 cannot be a King because from [3] one King must be between two Jacks and [4] prevents this situation. From previous reasoning, card 6 cannot be a Queen. Card 6 cannot be the Ace, from [6]. So card 6 must be a Jack. But then [3] and [7] cannot both be satisfied, so situation (b) is impossible.

If only card 6 is a Queen, then either of the following partial arrangements must occur.

```
      X                X
  X X X            X X K
    K Q K            X Q X
      X                K
```

Statements [3] and [4] cannot both be satisfied in the first arrangement and [3] cannot be satisfied in the second arrangement, so situation (c) is impossible.

So situation (d) is the correct one: only card 4 is a Queen. Then, from [2], cards 1 and 6 are Kings. Then, from [3], cards 5 and 7 are Jacks. So the following partial arrangement must occur (*J* represents a Jack):

```
      K
  X X Q
    J K J
      X
```

If cards 2 and 3 are both Kings, as required by [7], then card 8 is the Ace, as required by [5]. But [6] rules out this situation. So card 8 is the King that borders on a King required by [7].

If card 2 is an Ace then card 3 cannot be a Queen (from [2]), a King (from [6]), a Jack (from [4]), or an Ace (from [5]). Since [8] cannot be satisfied for card 3, card 2 is not an Ace. From [5], ***card* 3 *must be the only Ace.***

To complete the picture, card 2 must be a Jack, from [2], [5], and [6].

```
      K
  J A Q
    J K J
      K
```

27 / The Murderess

Because each statement refers to a different woman, the innocent one did not make all three statements; otherwise, she would have spoken of herself, contradicting [I]. So the innocent one made either one statement or two statements, from [II].

If the innocent one made only one statement, then only that statement is true and the other two statements are false, from [III]. But this situation is impossible, because if any two of these statements are false, then the remaining one has to be false, as follows:

(a) If [1] and [2] are false, then Anna is the accomplice and Babs is the murderess. So Cora must be the innocent one, making [3] false.

(b) If [1] and [3] are false, then Anna is the accomplice and Cora is the innocent one. So Babs must be the murderess, making [2] false.

(c) If [2] and [3] are false, then Babs is the murderess and Cora is the innocent one. So Anna must be the accomplice, making [1] false.

So the innocent woman made two statements. From [I], the two true statements were made by the only woman not referred to in these two statements:

(d) If statements [2] and [3] are true, they were made by Anna. Then Anna is the innocent one. But [1], being false, identifies Anna as the accomplice. This situation is impossible.

(e) If statements [1] and [3] are true, they were made by Babs. Then Babs is the innocent one. But [2], being false, identifies Babs as the murderess. This situation is impossible.

(f) So statements [1] and [2] are true, and were therefore made by Cora. Then Cora is the innocent one. The falsity of [3] is consistent with this conclusion. Since Cora is the innocent one and [1] is true, Babs is the accomplice. Then ***Anna is the murderess.*** [2], being true, is consistent with this conclusion.

28 / Conversation Stopper

Let a = the number of people in each group at 8 P.M. Then, from [1], there were $5a$ people at the party at 8 P.M. Let b = the number of people in each group at 9 P.M. Then, from [2], there were $4b$ people at the party at 9 P.M. Since only two people arrived between 8 P.M. and 9 P.M., from [1] and [2], $5a + 2 = 4b$.

Let c = the number of people in each group at 10 P.M. Then, from [3], there were $3c$ people at the party at 10 P.M. Since only two people arrived between 9 P.M. and 10 P.M., from [2] and [3], $4b + 2 = 3c$.

Let d = the number of people in each group at 11 P.M. Then, from [4], there were $2d$ people at the party at 11 P.M. Since only two people arrived between 10 P.M. and 11 P.M., from [3] and [4], $3c + 2 = 2d$.

Trial and error yields the following values for a, b, and c in the first and second equations (a cannot be greater than 20, from [1]).

$5a + 2 = 4b$

a	b
2	3
6	8
10	13
14	18
18	23

$4b + 2 = 3c$

b	c
1	2
4	6
7	10
10	14
13	18
16	22
19	26
22	30

Since b must have the same value in both equations, $b = 13$; then $a = 10$ and $c = 18$. Since $c = 18$, $d = 28$ in the third equation.

So at 8 P.M. there were 50 people, at 9 P.M. there were 52 people, at 10 P.M. there were 54 people, and at 11 P.M. there were 56 people.

From [1], [5], and [6], if Amelia were the woman to arrive an hour after her husband the number of people present at 8 P.M. would be 49. From [2], [5], and [6], if Brenda were the woman to arrive an hour after her husband the number of people present at 9 P.M. would be 51. From [3], [5], and [6], if Cheryl were the woman to arrive an hour after her husband the number of people present at 10 P.M. would be 53. From [4], [5], and [6], if Denise were the woman to arrive an hour after her husband the number of people present at 11 P.M. would be 55.

Of the four numbers of people—49, 51, 53, and 55—53 people is the only number of people not divisible into smaller groups containing an equal number of people (the number of people in each group must be at least two for conversation to take place).

So, from [3] and [6], ***Cheryl is the woman who suspected her husband of infidelity.***

29 / The Constant Winner

From [2], there are sixteen ways the four players could have chosen to take chips from a pile. These ways are listed below. Using [1], suppose two chips are in a pile, then four chips, then six chips, then eight chips, and then ten chips. Using [3] and [4], record the winner of each game for each combination until a different winner occurs for a given combination. The winners are recorded below alongside the combinations. Combination nine: 1, 2, 2, 1 is the only combination yielding the same winner, Don, for each game. For all other even numbers of chips, ***Don will always win*** with this combination.

	ABE	BEN	CAL	DON	2 CHIPS	4 CHIPS	6 CHIPS	8 CHIPS	10 CHIPS
1.	1	1	1	1	Ben	Don	—	—	—
2.	2	1	1	1	Ben	Cal	—	—	—
3.	1	2	1	1	Cal	Cal	Abe	—	—
4.	1	1	2	1	Ben	Cal	—	—	—
5.	1	1	1	2	Ben	Abe	—	—	—
6.	2	2	1	1	Abe	Ben	—	—	—
7.	2	1	2	1	Abe	Don	—	—	—
8.	2	1	1	2	Abe	Cal	—	—	—
9.	1	2	2	1	Don	Don	Don	Don	Don
10.	1	2	1	2	Cal	Cal	Don	—	—
11.	1	1	2	2	Ben	Cal	—	—	—
12.	1	2	2	2	Abe	Abe	Abe	Abe	Ben
13.	2	1	2	2	Abe	Ben	—	—	—
14.	2	2	1	2	Abe	Ben	—	—	—
15.	2	2	2	1	Abe	Ben	—	—	—
16.	2	2	2	2	Abe	Ben	—	—	—

30 / The Woman Hector Will Marry

If Bernice nags, then from [3] and [5] all three women nag. If Claudia nags, then from [5] Annette nags. Annette may be the only woman who nags. So, from [7], the possible combinations for nagging women are (*B* represents Bernice, *C* represents Claudia, and *A* represents Annette):

BCA, *CA*, and *A*.

From [1] and [4], either both Annette and Bernice talk a lot or neither of them talk a lot. Claudia may or may not talk a lot. So, from [7], the possible combinations for women who talk a lot are:

BCA, *AB*, and *C*.

If Annette is stubborn, then from [2] Claudia is stubborn. If Bernice is stubborn, then from [6] Claudia is not stubborn. So Annette and Bernice cannot both be stubborn. Claudia may be the only woman who is stubborn or Bernice may be the only woman who is stubborn. So, from [7], the possible combinations for women who are stubborn are:

AC, *C*, and *B*.

From [8], if the two women mentioned possess only one of the same unfortunate traits, then it is impossible to satisfy both [7] and [9].

If the two women mentioned in [8] possess either two or three of the same unfortunate traits, then the two women cannot be Bernice and Claudia because Annette will have more than one trait, contradicting [9]. The two women cannot be Annette and Bernice because then Claudia will have more than one trait, contradicting [9]. So Annette and Claudia are the two women and ***Bernice must be the woman Hector will marry.***

The only possible distribution of traits is:

Nags	*CA*
Talks a lot	*BCA*
Is stubborn	*AC*

31 / Father and Son

Let:

a represent the number of shares Arnold bought
b represent the number of shares Barton bought
c represent the number of shares Claude bought
d represent the number of shares Dennis bought

Then the equation that represents the total money spent is:

$$3a + 4b + 6c + 8d = 161 \quad \text{(from [1] and [4])}.$$

From [1] and [2], if Arnold is the father he bought 24 shares; and, from [1] and [3], if Barton is his son he bought 6 shares, etc. There are twelve possibilities, which may be tabulated as follows:

	FATHER (*spent* \$72)	SON (*spent* \$24)
I	$a = 24$	$b = 6$
II	$a = 24$	$c = 4$
III	$a = 24$	$d = 3$
IV	$b = 18$	$a = 8$
V	$b = 18$	$c = 4$
VI	$b = 18$	$d = 3$
VII	$c = 12$	$a = 8$
VIII	$c = 12$	$b = 6$
IX	$c = 12$	$d = 3$
X	$d = 9$	$a = 8$
XI	$d = 9$	$b = 6$
XII	$d = 9$	$c = 4$

NOTE THAT: A. *each of the letters a, b, c, and d represents a positive whole number, and* B. *if a number divides exactly into four of the five terms in the equation, then that number must divide exactly into the fifth term.*

From [B], above, a does not equal 24 or 8, because 161 is not divisible by 2; b cannot equal 18 if d equals 3, nor can d equal 9 if b equals 6, because 161 is not divisible by 3. So cases I, II, III, IV, VI, VII, X, and XI are eliminated.

If $d = 9$ and $c = 4$, then $3a + 4b = 65$; it follows that a or b is greater than 9, which contradicts [2]. If $c = 12$ and $b = 6$, then $3a + 8d = 65$; it follows that a or d is less than 6, which contradicts [3]. So cases VIII and XII are eliminated.

If $b = 18$ and $c = 4$, then $3a + 8d = 65$. $3a$ has to be an odd number because $8d$ is even and 65 is odd (an even number times any whole number is even, and an even number plus an odd number equals an odd number).

Then a has to be an odd number between 4 and 18 (an odd number times an odd number is odd). The only possibility allowing a whole number value for d is $a = 11$; but this means $d = 4$, which contradicts [3]. So case V is eliminated.

Case IX is the only remaining possibility, so ***Claude is Dennis' father.***

Further analysis yields two sets of possible values for a, b, c, and d. If $c = 12$ and $d = 3$, then $3a + 4b = 65$. a has to be an odd number between 12 and 3, by reasoning as was done earlier. The only possibilities allowing whole number values for b are $a = 7$ and $a = 11$. The two sets of possible values are:

$a = 7$	$a = 11$
$b = 11$	$b = 8$
$c = 12$	$c = 12$
$d = 3$	$d = 3$

32 / The Race

If two of the statements [1], [3], [5], and [7] are true, then a third one has to be true; so of statements [1], [3], [5], and [7] it is impossible for exactly two of them to be false. If three of the statements [2], [4], [6], and [8] are false, then the fourth has to be false; so of statements [2], [4], [6], and [8] it is impossible for exactly three of them to be false.

Therefore, either one, three, or four statements of [1], [3], [5], and [7] are false (at least one has to be false), and either none, one, two, or four statements of [2], [4], [6], and [8] are false.

From [I], six statements are false altogether. There is only one way to get a total of six by adding two of these numbers of possible false statements: four plus two. Therefore, statements [1], [3], [5], and [7] are all false and two of statements [2], [4], [6], and [8] are false.

If [2] is false, Alan won the race, which contradicts [II]. So [2] is true. Then, either: [2] and [4] are true, [2] and [6] are true, or [2] and [8] are true.

If [2] and [4] are true, then [6] and [8] are false, and the order in which the men finished the race is *BACD* (the letters are the first letters of the names of the four men). But this order contradicts [5] as false, so it is not the correct one.

If [2] and [8] are true, then [4] and [6] are false, and the order in which the men finished the race is *DBCA*. But this order contradicts [3] as false, so it is not the correct one.

So [2] and [6] must be true, which implies that [4] and [8] are false, and the order in which the men finished the race is *CBAD*. Therefore, ***Clay won the race.***

33 / Murder by Profession

From [1], the doctor in the bar was either Alex, Bell, or Dean. From [3] and [4], the doctor and lawyer in the bar were both from one family, and the killer and victim were both from the other family. Choosing a doctor and a lawyer from one family (there are four possible combinations to choose) and a killer and a victim from the other family (using [4]) will leave two remaining people who attended a movie.

If Bell and Cass were in the bar (combination I), then Dean and Earl were on the beach (from [4]), and Alex and Faye were at the movie. But, from [6a], neither Alex nor Faye can be the ex-spouse of Bell or Cass (Alex is from the same family as Bell and Cass, and Faye is of the same sex as Bell and Cass). So Bell and Cass were not in the bar.

Conversely, if Dean and Earl were in the bar (combination II), then Bell and Cass were on the beach (from [4]), and Alex and Faye were, again, at the movie. But, from [6a], neither Alex nor Faye can be the ex-spouse of Dean or Earl (Faye is from the same family as Dean and Earl, and Alex is of the same sex as Dean and Earl). So Dean and Earl were not in the bar.

From the previous reasoning, either Alex and Cass were in the bar (combination III) or Dean and Faye were in the bar (combination IV). If Alex and Cass were in the bar, Earl cannot be the victim, from [5], and the victim and the killer were of opposite sex, from [4]. If Dean and Faye were in the bar, Bell cannot be the victim, from [5], and the victim and the killer were of opposite sex, from [4].

So combinations III and IV lead to six possible distributions (*D* denotes doctor and *L* denotes lawyer):

Role	(a)	(b)	(c)	(d)	(e)	(f)
Doctor in bar	Alex (*D*)	Alex (*D*)	Alex (*D*)	Dean (*D*)	Dean (*D*)	Dean (*D*)
Lawyer in bar	Cass (*L*)	Cass (*L*)	Cass (*L*)	Faye (*L*)	Faye (*L*)	Faye (*L*)
Victim	Dean (*D*)	Faye (*L*)	Faye (*L*)	Alex (*D*)	Cass (*L*)	Alex (*D*)
Killer	Faye (*L*)	Dean (*D*)	Earl (*L*)	Cass (*L*)	Alex (*D*)	Bell (*D*)
At movie	Bell (*D*) Earl (*L*)	Bell (*D*) Earl (*L*)	Bell (*D*) Dean (*D*)	Bell (*D*) Earl (*L*)	Bell (*D*) Earl (*L*)	Cass (*L*) Earl (*L*)

Possibilities (a), (b), (d), and (e) contradict [2], so are eliminated.

In (c), Dean must be the ex-spouse of Cass, from [6a]. If that is true, then Bell and Alex must be former roommates, which contradicts [6b] (Bell and Alex are not of the same sex). So possibility (c) is eliminated.

Since (f) is the only remaining possibility, ***Bell must be the killer.*** Checking out [6], Cass must be the ex-spouse of Dean, and Earl must be the former roommate of Dean.

34 / Six G's

$F \times ABCDE = GGGGGG.$

$F \times ABCDE = G \times 111111.$

Of the numbers 2 through 9, 111111 is divisible exactly by only 3 and 7.

$F \times ABCDE = G \times 3 \times 7 \times 5291.$

If G is a multiple of F, then $ABCDE$ would be a number containing the same digit 6 times. So G is not a multiple of F.

Then: (a) F does not equal zero, otherwise G would equal zero and, therefore, would be a multiple of F.

(b) F does not equal 1, otherwise G would be a multiple of F.

(c) F does not equal 2, otherwise G would have to be a multiple of 2 (for an exact division) and, therefore, a multiple of F.

(d) F does not equal 4, otherwise G would have to be a multiple of 4 (for an exact division) and, therefore, a multiple of F.

(e) F does not equal 8, otherwise G would have to be 8 also (for an exact division) and, therefore, a multiple of F.

(f) F does not equal 5, otherwise G would have to be 5 also (for an exact division) and, therefore, a multiple of F.

(g) If $F = 3$, then: $ABCDE = G \times 7 \times 5291 = G \times 37037$. The presence of a zero in 37037 indicates that the product of any single digit times this number will result in duplicate digits for $ABCDE$. So F does not equal 3.

(h) If $F = 6$, then: $ABCDE \times 2 = G \times 7 \times 5291 = G \times 37037$. G, then, must be a multiple (M) of 2, that is, $G/2 = M$. Then: $ABCDE = M \times 37037$. By the reasoning in (g), F does not equal 6.

(i) If $F = 9$, then: $ABCDE \times 3 = G \times 7 \times 5291 = G \times 37037$. So G must be a multiple (M) of 3, that is, $G/3 = M$. Then: $ABCDE = M \times 37037$. By the reasoning in (g), F does not equal 9.

(j) So $F = 7$. Then: $ABCDE = G \times 3 \times 5291 = G \times 15873$. Since there are seven different digits involved, G does not equal 1, 5, or 7. Since $ABCDE$ contains only five digits, G does not equal 8 or 9. Since F does not equal 0, G does not equal zero. So G equals 2, 3, 4, or 6.

The four possibilities are:

$$F = 7,\ G = 2,\ ABCDE = 31746$$
$$F = 7,\ G = 3,\ ABCDE = 47619$$
$$F = 7,\ G = 4,\ ABCDE = 63492$$
$$F = 7,\ G = 6,\ ABCDE = 95238$$

Only the last one of these possibilities results in seven different digits. The multiplication, then, is

$$\begin{array}{r} 9\ 5\ 2\ 3\ 8 \\ \times \qquad 7 \\ \hline 6\ 6\ 6\ 6\ 6\ 6 \end{array}$$

and *G* ***represents 6.***

35 / Two or Three

From [II], if a player *can* win, he *must* win; if a player *can* force a tie (assuming he cannot win), then he *must* force a tie.

From [2] and [3]:

(a) If a pile contains only one coin, then the game must end in a tie because neither player can draw from the pile.

(b) Drawing from two coins, a player loses because he must take the two coins.

(c) If a pile contains three coins, then the game must end in a tie. If a player takes three coins he loses. If he takes two coins the other player cannot draw.

(d) Drawing from four coins, a player wins by taking two coins, thus putting the other player in the losing position of drawing from two coins. If he takes three coins, then the game ends in a tie.

(e) Drawing from five coins, a player wins if he is able to leave a number of coins that represents a losing position for the other player. Accordingly, he takes three coins, putting the other player in the losing position of drawing from two coins.

(f) If a pile contains six coins, then the game must end in a tie. A player draws three coins, leaving three coins which represents a tieing position. If a player draws two coins, he gives the other player the winning position of drawing from four coins.

Reasoning in this manner, one finds that drawings from two, seven, and twelve coins are losing positions; drawings from four, five, nine, and ten coins are winning positions; and drawings from one, three, six, eight, and eleven coins are tieing positions. The following tables summarize how these three sets of drawings can be losing, winning, and tieing positions, respectively.

FROM A LOSING POSITION OF	IF A PLAYER DRAWS	HE LEAVES A WINNING POSITION OF
2	2	0
7	{2, 3}	{5, 4}
12	{2, 3}	{10, 9}

FROM A WINNING POSITION OF	DRAW (2 *or* 3)	TO LEAVE A LOSING POSITION OF
4	2	2
5	3	2
9	2	7
10	3	7

FROM A TIEING POSITION OF	DRAW (2 *or* 3)	TO LEAVE A TIEING POSITION OF
1	—	1
3	2	1
6	3	3
8	2	6
11	3	8

From [1], there are twelve coins. Since drawing from twelve coins is a losing position, whoever goes first must lose. Since Armand goes first, from [I], Armand must lose. So ***Buford must win.***

36 / A Week in Cantonville

From [1], [5], and [8], the supermarket is open on Monday, Tuesday, Friday, and Saturday.

From [2], [5], and [7], the department store is not closed on both Saturday and Monday.

From [3], [5], and [6], the bank is not closed on both Saturday and Monday.

From [9] and previous reasoning, either:

(A) the bank is the only place closed on Saturday and the department store is the only place closed on Monday, or

(B) the department store is the only place closed on Saturday and the bank is the only place closed on Monday.

If (A) is correct, then from [2], [5], and [7] the days on which the department store is open can be determined (*C* stands for closed, *O* for open):

	SUN	MON	TUES	WED	THURS	FRI	SAT
Bank	*C*	*O*					*C*
Department store	*C*	*C*	*O*		*C*	*O*	*O*
Supermarket	*C*	*O*	*O*				*O*

This situation is impossible because it contradicts [4].

So (B) is correct. Using [2], [5], and [7] again, the following table results:

	SUN	MON	TUES	WED	THURS	FRI	SAT
Bank	*C*	*C*					*O*
Department store	*C*	*O*	*O*	*C*	*O*	*O*	*C*
Supermarket	*C*	*O*	*O*			*O*	*O*

From [3] and [6], the table can be filled in further:

	SUN	MON	TUES	WED	THURS	FRI	SAT
Bank	*C*	*C*	*O*	*C*	*O*	*C*	*O*
Department store	*C*	*O*	*O*	*C*	*O*	*O*	*C*
Supermarket	*C*	*O*	*O*			*O*	*O*

From the above table, ***Tuesday must be the day all three places are open.***

To complete the table: from [1] and from the fact that all three places are open on only one day each week, the supermarket must be open on Wednesday.

37 / The Book Shelf

Using [4], let:

c = the width of a catalogue
d = the width of a dictionary
e = the width of an encyclopedia
x = the width of the shelf

Then, according to each assistant's statement:

(A) Astor: $2c + 3d + 3e = x$
(B) Brice: $4c + 3d + 2e = x$
(C) Crane: $4c + 4d + 3e = x$

Subtracting C from A: $d + 2c = 0$, and $d = -2c$, which is impossible.
Subtracting C from B: $d + e = 0$, and $d = -e$, which is impossible.
Subtracting B from A: $e - 2c = 0$, and $e = 2c$, which *is* possible.

Since the use of C with each of A and B results in an impossible situation, Mrs. Crane was incorrect. So, from [1], equations A and B are correct, and $e = 2c$.

From [3], if 15 encyclopedias can fill the shelf exactly, then so can 30 catalogues. Therefore, from [2], the encyclopedias cannot fill the shelf.

If 15 dictionaries can fill the shelf, then using A (or B) and $e = 2c$:

$$\begin{aligned} 2c + 3d + 3e &= x \\ e + 3d + 3e &= x \\ 3d + 4e &= x \\ 3d + 4e &= 15d \\ 4e &= 12d \\ e &= 3d \end{aligned}$$

From [3], if 15 dictionaries can fill the shelf exactly, then so can 5 encyclopedias. Therefore, from [2], the dictionaries cannot fill the shelf.

So the catalogues can fill the shelf.

Using A (or B) and $e = 2c$:

$$
\begin{aligned}
2c + 3d + 3e &= x \\
2c + 3d + 6c &= x \\
3d + 8c &= x \\
3d + 8c &= 15c \\
3d &= 7c \\
d &= 2\tfrac{1}{3}c
\end{aligned}
$$

If 15 catalogues can fill the shelf exactly, then the encyclopedias cannot; using $e = 2c$, $7\frac{1}{2}$ encyclopedias would fill the same space. The dictionaries cannot fill the shelf exactly; using $d = 2\frac{1}{3}c$, $6\frac{3}{7}$ dictionaries would fill the same space.

38 / The Hostess

Clubs is not trump; otherwise, from [1] and [4], Alma would have led more than once, which contradicts [3].

Hearts is not trump; otherwise, from [2] and [4], the hostess would have won more than once, which contradicts [5].

From [1], no one followed Alma's lead of a club, indicating no one else had a club; yet, from [4], a club was played at each trick. So Alma must have had three clubs. Since a trump won each of the last three tricks and clubs cannot be trump, Alma did not win any of the tricks. From [5], each of the other three women won one trick. Therefore, each of the other three women had one trump.

Spades is not trump; otherwise, no one had three red cards which contradicts [6].

So diamonds is trump.

Then, from [1], Bess won at trick number eleven and led at trick number twelve.

From [2], the hostess won the twelfth trick (with a trump) and led a heart; so, from [4], hearts was not led at trick number twelve.

Diamonds could not have been led at trick number twelve because Bess would have won more than one trick which contradicts [5] (having won at trick number eleven, she would also have won at trick number twelve, from [4]).

Clubs could not have been led at trick number twelve because Alma had all the clubs and, from [3], she led only once (at trick number eleven, from [1]).

So spades was led by Bess at trick number twelve. The table below records the suits known to be played by each woman so far.

	ALMA	BESS	CLEO	DINA
Eleventh trick:	club (led)	diamond (won)	heart	spade
Twelfth trick:	club	spade (led)		
Thirteenth trick:	club			

If Bess led a spade at trick number twelve, then either Cleo played a diamond (trump) or Dina did, from [5]. If Cleo did, she must have been the hostess, from [2]. But, from [6], the hostess' partner held three red cards, and none of the other women could have been the hostess' partner (Alma had all clubs, Bess led a spade at trick number twelve, and Dina played a spade at trick number eleven). So Cleo did not play a diamond (trump) when Bess led a spade at trick number twelve.

So Dina must have played a diamond (trump) when Bess led a spade at trick number twelve. Then, from [2], ***Dina must have been the hostess.***

The analysis may be continued. From [2], Dina led a heart at trick number thirteen. So the table may be filled in further.

	ALMA	BESS	CLEO	DINA
Eleventh trick:	club (led)	diamond (won)	heart	spade
Twelfth trick:	club	spade (led)		diamond (won)
Thirteenth trick:	club			heart (led)

Then, from [4], Cleo played a heart at trick number twelve. From [5], Cleo played a diamond (trump) at trick number thirteen; then, from [4], Bess played a spade at trick number thirteen. The completed table is shown below.

	ALMA	BESS	CLEO	DINA
Eleventh trick:	club (led)	diamond (won)	heart	spade
Twelfth trick:	club	spade (led)	heart	diamond (won)
Thirteenth trick:	club	spade	diamond (won)	heart (led)

39 / The Rectangular Table

From [7], the killer and the victim were of opposite sex. From [3], the victim and the killer each sat across from a member of the opposite sex. Then, from [1] and [2], one of the two partial seating arrangements below must be correct (*M* stands for man and *W* stands for woman).

	I			II	
	M — —			*W* — —	
M		*W*	*W*		*M*
	W — —			*M* — —	

The person seated next to the victim was either a man or a woman. From [3], the person sitting opposite this latter person was a member of the opposite sex. If a man sat next to the victim, it would not be possible to complete either seating arrangement and satisfy [4] at the same time: from [4], at least one man sat between two women, so Ia, IIa, and IIb (see below) are not possible; from [4], at most one man sat between two women, so Ib is not possible (see below).

	Ia			Ib			IIa			IIb	
	M W W			*M W M*			*W W W*			*W W M*	
M		*W*	*M*		*W*	*W*		*M*	*W*		*M*
	W M M			*W M W*			*M M M*			*M M W*	

So the person who sat next to the victim must have been a woman: and, from [3], a man sat across from her. In arrangement I, if a woman sat next to this woman, it would not be possible to complete this seating arrangement and satisfy [4] at the same time: from [4], at least one man sat between two women, so Ic (see below) is not possible. In arrangement II, if a man sat next to this woman it would not be possible to complete this seating arrangement and satisfy [4] at the same time: from [4], at most one man sat between two women, so IIc (see below) is not possible.

	Ic			Id			IIc			IId	
	M M M			*M M W*			*W M W*			*W M M*	
M		*W*	*M*		*W*	*W*		*M*	*W*		*M*
	W W W			*W W M*			*M W M*			*M W W*	

So either Id or IId represents the correct seating arrangement. In each case, [3] and [4] establish the positions of the host and hostess, [5] identifies the women sitting next to the host, and [3] identifies the men sitting across from these women. So one of the four partial seating arrangements below must be correct:

	□	Barry	Hostess	
Nathan		Id_1		Natalie
	□	Barbara	Host	

	Hostess	Barry	□	
Natalie		IId_1		Nathan
	Host	Barbara	□	

	□	Nathan	Hostess	
Barry		Id_2		Barbara
	□	Natalie	Host	

	Hostess	Nathan	□	
Barbara		IId_2		Barry
	Host	Natalie	□	

From [6], Id_1 and IId_1 are not possible. Since only one man and one woman remain unaccounted for, arrangements Id_2 and IId_2 can now be completed.

	Samuel	Nathan	Hostess	
Barry		Id_2		Barbara
	Samantha	Natalie	Host	

	Hostess	Nathan	Samuel	
Barbara		IId_2		Barry
	Host	Natalie	Samantha	

From [1], [2], and [7], Id_2 is not possible because Barry and Samantha are brother and sister (Barry is the host's brother and Samantha is the host's sister). So IId_2 is the correct seating arrangement.

Therefore, from [1] and [2], ***the host was killed by his brother's wife, Barbara.***

40 / Deanna's Sister

Let P = number of Deanna's pennies
N = number of Deanna's nickels
Q = number of Deanna's quarters
T = total cost of candy in cents
a = number of Althea's candies
b = number of Blythe's candies
c = number of Carrie's candies
d = cost of one favor in cents
F = number of favors

All the numbers involved are positive whole numbers.

From [1]: (1a) $P + N + Q = 13$

(1b) $P + 5N + 25Q = T$

From [2]: (2) $2a + 3b + 6c = T$

From [3]: (3) a, b, c: all different; all greater than 1

From [4]: (4) $2a = 3b$, $2a = 6c$, or $3b = 6c$

From [5]: (5) $F \times d = 480$

From [6]: (6) $a + b + c = F$

From [7], the problem can be restated as:

(7) which is greatest: a, b, or c?

There are six equations in nine unknowns, the fourth equation being one of three possibilities. There are too many equations to be able to use algebra alone in solving the problem; so additional characteristics of the numbers involved, besides the fact that they are all positive whole numbers, must be sought.

Now, the sum of two odd numbers is always even
the sum of two even numbers is always even
the sum of one odd and one even number is always odd

Also, the product of two odd numbers is always odd
the product of two even numbers is always even
the product of one odd and one even number is always even

From this information, in equation (1a), either all three of P, N, and Q are odd or only one of them is odd; T in equation (1b) is odd in either case. Then b in equation (2) is odd. Then, in (4), $2a$ cannot equal $3b$ because $2a$ is even and $3b$ is odd. Neither can $3b$ equal $6c$ because $6c$ is even and $3b$ is odd. So $2a = 6c$. (At this point it is known that c is not the greatest because a has to be larger than c.) Dividing by 2, $a = 3c$. Substituting for a in (6), $b + 4c = F$.

Now b is still odd; so, in $b + 4c = F$, F is odd. In (5), 480 is the product of two numbers, one of which is odd (F) and one of which is even (d). The only odd-number values possible for F in this product are 1, 3, 5, and 15. 1 and 3 are not possible values for F because b and c must both be positive whole numbers. 5 is not possible from [3] (b and c cannot equal 1). So F must equal 15.

Then $b + 4c = 15$, and c cannot equal more than 3 or less than 1. From [3], c does not equal 1 or 3. So c must equal 2. Then $b = 7$. Recalling that $a = 3c$, then $a = 6$. So b is greatest. Therefore, from [7], ***Blythe is Deanna's sister.***

All other values can be found as follows. If $F = 15$, then, from [5], $d = 32$. If $a = 6$, $b = 7$, and $c = 2$, then, from [2], $T = 45$. Subtracting

(1a) from (1b) gives $4N + 24Q = 32$. Dividing by 4, $N + 6Q = 8$. Q is neither greater than 1 (otherwise N would be negative) nor less than 1 (there were three denominations, from [1]), so $Q = 1$. Then $N = 2$. Then, from (1a), $P = 10$.

41 / The Cube

There are three possible arrangements of the figures on the faces of the cube, if the owner's statement is disregarded. Two of these arrangements are excluded by the owner's statement.

Either any one of the figures occurs only once on the cube or it occurs twice. If one figure is chosen, reasoning can then proceed. Which figure is it convenient to choose? Since both the ○ and the ● occur with four different figures, then, if one of them occurs only once, the figures on four other faces can be derived at once.

If the ● is chosen, then there are two possibilities, as previously mentioned: either ● occurs once or it occurs twice.

If ● occurs once, then from the second view: 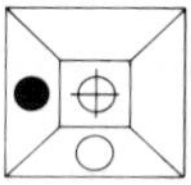 Then from the third view: 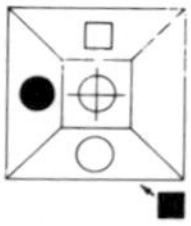 Finally, from the first view: either 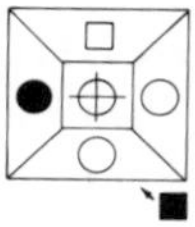

I

or 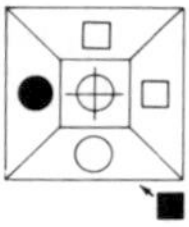

II

If ● occurs twice, then every other figure occurs only once. From the second view: 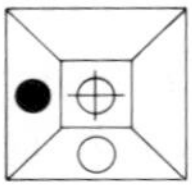 The ○ pictured in the first view is the same ○ pictured in the second view. So from the first view: 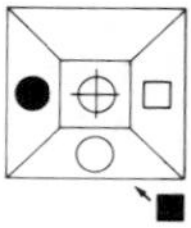

Then from the assumption that ● occurs twice:

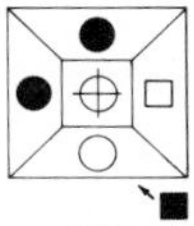

III

The third view checks out the assumption.

The figure on the bottom face in each view is recorded below for each of the three possible arrangements. Also recorded is the figure that occurs twice in each arrangement.

BOTTOM FIGURES

		View 1	*View 2*	*View 3*	*Repeated figure*
I	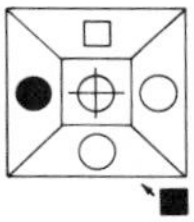	●	■	⊕	○
II	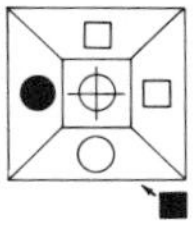	□	■	⊕	□
III	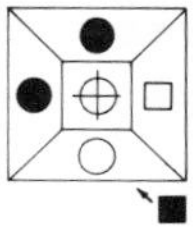	●	■	⊕	●

From the owner's statement, arrangements II and III are not possible. So arrangement I is correct, and ○ ***occurs twice.***

42 / The Club Trick

From [7], four different suits were led.

At the club trick: from [2], everyone played a club.

At the diamond trick: from [4], diamonds were led first, so three diamonds were played; Deb must have been the only one not to follow suit and she must have played a spade (from [2], she followed suit at the club trick).

At the spade trick: only two spades could have been played (from [1]), so Bea must have played one of her hearts (from [2], she followed suit at the club trick and, from [4], she must have followed suit at the diamond trick); Cyd must have played a heart or a diamond (from [2], she followed suit at the club trick).

At the heart trick: Ada and Bea must have each played a heart (each was able to play one heart at this point, from previous reasoning), Deb must have played a spade (from [2], she followed suit at the club trick), and Cyd must have played a heart or a diamond (from [2], she followed suit at the club trick).

From previous reasoning and from [3], only two hearts were played at the heart trick. So Cyd played a diamond at the heart trick; then she must have played a heart at the spade trick.

The above conclusions concerning the suit played by each player at each trick are summarized in the table below:

	ADA	BEA	CYD	DEB
club trick:	club	club	club	club
diamond trick:	diamond	diamond	diamond	spade
heart trick:	heart	heart	diamond	spade
spade trick:	spade	heart	heart	spade

Diamonds were led first, from [4]. Clubs and hearts were not led second and third in either order because Cyd would not have been able to play a heart when spades were led (she must follow suit if she can). For the same reason, hearts were not led second.

Also, either the club trick followed the spade trick or the spade trick followed the club trick (from [1], Deb had only these two suits; from [5], she had to lead at one of these two tricks; from [8], she had to win the other of these two tricks).

Therefore, the order of suits led was either:

I		II
diamonds		diamonds
spades	*or*	clubs
clubs		spades
hearts		hearts

From [5] and [6], the player who led the diamond won the heart. Only Ada and Bea played a heart at the heart trick; so this player is either Ada or Bea, from [8]. Then Ada or Bea led the diamond and the other of the two led the heart.

From [1] and [5], Deb must have led the spade (Ada and Bea led as previously indicated, and Cyd had no spades).

Therefore, ***Cyd led the club.***

The analysis may be continued. Since the spade trick could only have been won by Ada (from [1], [5], [8], and the fact that Deb led the spade), clubs did not follow spades (because Cyd led the club and could not have won the spade). So order II must be correct. Then hearts followed spades and Ada led the heart. Then Bea led the diamond.

The play of the last four tricks is shown below:

	ADA	BEA	CYD	DEB
diamond trick:	followed suit	led	won	played a spade
club trick:	followed suit	followed suit	led	won
spade trick:	won	played a heart	played a heart	led
heart trick:	led	won	played a diamond	played a spade

43 / Twelve C's

$K + C = C$ implies $K = 0$.

$$\begin{array}{c} K \\ C \\ \hline C \end{array}$$

A does not equal zero, otherwise the second partial product would equal zero.

A does not equal 1, otherwise the second partial product would be a duplicate of the first row (*ABCDEFGH*).

$K = 0$ implies E does not equal 0, which implies B is less than 9 (nothing was carried from $E + B$).

$$\begin{array}{c} E \\ B \\ \hline C \end{array}$$

B is less than 9 implies *A* is less than 3.

$$\begin{array}{r} A \\ \hline A \\ \hline B \qquad\quad \\ \hline \end{array}$$

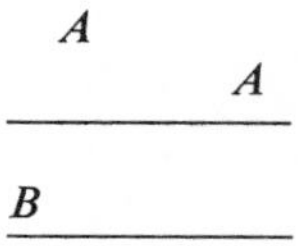

A does not equal 0 or 1 and *A* is less than 3 imply *A* = 2.

A = 2 and *B* is less than 9 imply *A* × *B* is less than or equal to 16.

A × *B* is less than or equal to 16 implies that no more than 1 could have been carried to *A* × *A* (*A* × *C* cannot be more than 18, so no more than 2 could have been carried to *A* × *B*).

$$\begin{array}{r} A\ B\ C \\ \hline A \\ \hline B \qquad\qquad \\ \hline \end{array}$$

So either *A* × *A* = *B* or *A* × *A* + 1 = *B*. Then, since *A* = 2, either *B* = 4 or *B* = 5.

A = 2 implies *J* × *A* is less than or equal to 18; *B* = 4 or 5 implies *J* × *B* is less than or equal to 45. So *E* = 1 or *E* = 2.

$$\begin{array}{r} A\ B \\ \hline J \\ \hline E \qquad\quad \\ \end{array}$$

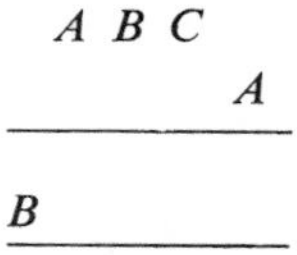

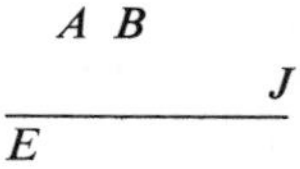

Then, since *A* = 2, *E* = 1.

E = 1 and *B* = 4 or 5 imply *C* = 5, 6, or 7;

$$\begin{array}{c} \overline{E} \\ B \\ \hline C \end{array}$$

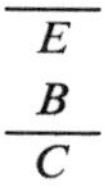

while *A* = 2 implies *C* is even.

$$\begin{array}{c} A \\ \hline K \\ C \end{array}$$

C = 5, 6, or 7 and *C* is even imply **C = 6.**
Then *G* = 5.

$$\begin{array}{c} \overline{G\ K} \\ E\ C \\ \hline C \end{array}$$

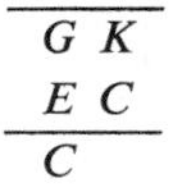

Then $B = 4$.

Then $F = 3$.

$$\begin{array}{c} \overline{A} \\ F \\ \hline C \end{array}$$

Then $H = 8$.

$$\begin{array}{cc} & H \\ A & \\ \hline & \\ C & \\ \hline \end{array}$$

Then $J = 7$.

$$\begin{array}{c} H \\ J \\ \hline C \\ \\ \hline \end{array}$$

Then $D = 9$.

The multiplication is shown below:

$$\begin{array}{r} 2\,4\,6\,9\,1\,3\,5\,8 \\ \times \qquad\qquad 2\,7 \\ \hline 1\,7\,2\,8\,3\,9\,5\,0\,6 \\ 4\,9\,3\,8\,2\,7\,1\,6 \\ \hline 6\,6\,6\,6\,6\,6\,6\,6\,6 \end{array}$$

44 / John's Ideal Woman

Among the women referred to in each of statements [1] through [4] is John's ideal woman. Not including her, statement [1] refers to two other women, statement [2] refers to one other woman, and statement [3] refers to one other woman. Since there are only four women altogether, one or more other women must be referred to more than once (as was the ideal woman). Since [1] refers to two other women, at least two other women are referred to in statements [1] through [4]. Since there are only three other women altogether, at most three other women are referred to.

Assume for the moment that four other women are referred to, two from statement [1], one from statement [2], and one from statement [3]. (Since either two or three other women are referred to, it will turn out that either one or two of these women are duplicates of another woman.) From statements [1] through [4], a table can be constructed as follows:

statements [1] through [4]		ideal woman:	blue-eyed	slender	tall	blonde
statement [1]	(i)	other woman:	blue-eyed	slender		
	(ii)	other woman:	blue-eyed	slender		
statement [2]	(iii)	other woman:			tall	blonde
statement [3]	(iv)	other woman:		slender	tall	

From [7], two women have different builds. Therefore, all the women are not slender; so (iii) is not slender. Since (i) and (ii) are known to be different women (they were mentioned in the same statement), and since (iii) is different from any of the others (she is the only one not slender), (iv) must be the same as (i) or (ii) (it does not make any difference which one). So the table now looks like this:

I (ideal)	blue-eyed	slender	tall	blonde
II (i and iv)	blue-eyed	slender	tall	
III (ii)	blue-eyed	slender		
IV (iii)		not slender	tall	blonde

From [4], II and III are not blonde and IV is not blue eyed. From [3], III is not tall. The table is now complete:

I	blue-eyed	slender	tall	blonde
II	blue-eyed	slender	tall	not blonde
III	blue-eyed	slender	not tall	not blonde
IV	not blue-eyed	not slender	tall	blonde

From [6], there are four possibilities:

	(a)	(b)	(c)	(d)
I	Betty	Carol		
II			Betty	Carol
III			Carol	Betty
IV	Carol	Betty		

From [5],(b) is not possible. From [8],(a) is not possible. There are again four possibilities:

	(c_1)	(c_2)	(d_1)	(d_2)
I	Doris	Adele	Doris	Adele
II	Betty	Betty	Carol	Carol
III	Carol	Carol	Betty	Betty
IV	Adele	Doris	Adele	Doris

From [7],(c_1) is not possible. From [5],(d_1) is not possible.

Since (c_2) and (d_2) are the only remaining possibilities, ***Adele is John's ideal woman.***

45 / The L-shaped Table

From [1], each man was seated between two women and each woman was seated between two men. This fact together with [4] indicates that Cain was married to either Fifi, Hera, or Joan.

(a) Abel is married to Babe. Assume Cain is married to Fifi. Then, from [1] and [4], Ivan must be married to Dido. Then Gene must be married to Joan. Then Ezra must be married to Hera.

(b) Abel is married to Babe. Assume Cain is married to Hera. Then, from [1] and [4], Ezra must be married to Joan. Then Gene must be married to Dido. Then Ivan must be married to Fifi.

(c) Abel is married to Babe. Assume Cain is married to Joan. Then, from [1] and [4], Ezra must be married to Hera. Then Gene must be married to Dido. Then Ivan must be married to Fifi.

So there are three possible sets of married partners. Using the first letter of each name for each person, they are as follows.

(a) *A—B, C—F, I—D, G—J, E—H*
(b) *A—B, C—H, E—J, G—D, I—F*
(c) *A—B, C—J, E—H, G—D, I—F*

From [1] and [6], the men and women were seated around the table in one of the three arrangements shown on page 108. (Statements [4] and [6] together indicate that "next to" refers to a person "sitting to the left or right around the table's perimeter.")

I	II	III
J	*A*	*G*
I *A*	*J* *B*	*F* *H*
B C	*C D*	*I J*
H *D*	*I* *E*	*E* *A*
G *F E*	*H* *G F*	*D* *C B*

From [2] and [8], the host and hostess and the two people sitting across from them are eliminated both as killer and victim in each arrangement. From [3] and [8], they are also eliminated as spouses of the victim and killer.

In each arrangement only one man and one woman sat across from each other. Therefore, from [2] and [3], the killer and the victim are of the same sex and their spouses are of the same sex. So the man and woman sitting across from each other are eliminated both as killer and victim in each arrangement; they are also eliminated as spouses of the victim and killer.

Then each of the four positions of killer, victim, killer's spouse, and victim's spouse must be contained in one of the following groupings.

I	II	III
—	—	—
— —	— —	*F* *H*
— *C*	*C D*	*I* —
H *D*	*I* *E*	— —
— — *E*	— *G F*	— *C* —

From [2] and [3], the four positions were occupied by two pairs of married couples, so two married couples must appear in a grouping.

In arrangement I, it is impossible for two married couples to occupy the four positions (neither (a), (b), nor (c) can occur). In arrangement II, (a), (b), and (c) give two possible sets of married couples for the four positions: *C—F*, *I—D*; *G—D*, *I—F*. In arrangement III, (b) provides one possible set of married couples for the four positions: *C—H*, *I—F*.

Remembering that the killer and victim were the same sex, that their spouses were the same sex, that the killer and victim sat across from each other, and that their spouses sat across from each other, one finds that each set of couples for arrangement II is impossible.

So (b) is the correct set of married partners and III is the correct seating arrangement.

From [7], Hera is not the killer. From [2] and [5], neither Cain nor Fifi is the killer. Therefore, ***Ivan is the killer.***

46 / The Tenth Trick

From the fact that the four hands contained four cards in each suit and from [6], only one trump was played at each trick. So, from [2] through [5], each of three players did not have a suit that was led at some point, and one player led a trump at some point (since four different suits were led). Further, since a trump must have won each trick, the player who led the trump must have had two trumps and must have led a trump at the last trick, having won the next-to-last trick. (If he had won an earlier trick than next-to-last he would have remained with the lead and have led twice, contradicting [2] through [5], which stipulate that four different players led.) Therefore, one player had two trumps, each of two other players had one trump, and one player had no trumps. From the distribution of the suits in the four hands, hearts or diamonds must be trump.

If diamonds were trump, Art held hand III, (from [2], Art led a diamond); while if hearts were trump, Bob held hand II (from [3], Bob led a heart). From [2], Art could not have held hand IV. From [3], Bob could not have held hand I. From [5], Dan could not have held hand II or hand III. There are, then, three possible combinations of holdings for each of the two possible trump suits:

	DIAMONDS TRUMP			HEARTS TRUMP		
	(a)	(b)	(c)	(d)	(e)	(f)
Hand I	Dan	Cab	Dan	Dan	Cab	Art
Hand II	Bob	Bob	Cab	Bob	Bob	Bob
Hand III	Art	Art	Art	Art	Art	Cab
Hand IV	Cab	Dan	Bob	Cab	Dan	Dan

Counting the number of trumps in each of the players' hands is equivalent to counting the number of tricks won by each player. For each of the six possible combinations of holdings, then,

(a) Art won 2, Bob won 1, Cab won 0, and Dan won 1
(b) Art won 2, Bob won 1, Cab won 1, and Dan won 0
(c) Art won 2, Bob won 0, Cab won 1, and Dan won 1
(d) Art won 1, Bob won 2, Cab won 1, and Dan won 0
(e) Art won 1, Bob won 2, Cab won 0, and Dan won 1
(f) Art won 0, Bob won 2, Cab won 1, and Dan won 1

Combinations (b), (c), (e), and (f) are eliminated by [7], and either (a) or (d) represents the correct holdings. Since (a) and (d) indicate the same holdings:

Art held hand	III	club – heart – diamond – diamond
Bob held hand	II	club – diamond – heart – heart
Cab held hand	IV	club – heart – spade – spade
Dan held hand	I	club – diamond – spade – spade

If diamonds were trump, then since one player did not win any of the last four tricks because he had no trumps, Cab led at trick ten (he had no trumps). But, from [4], Cab led a club. Since everyone had a club at this point, no one would be able to play a trump at the club lead. So Cab did not lead at trick ten and, therefore, diamonds were not trump.

If hearts were trump (as they must have been), then by similar reasoning Dan led at trick ten. Bob, having two hearts, led at trick thirteen. So either Art or Cab led at the eleventh trick, having won the tenth trick. Since, from [5], Dan led a spade, Cab was not in a position to play a trump at the spade lead at this point (he had two spades); so Art must have done so. Therefore, ***Art won the tenth trick.***

Below is the play of the remaining four tricks:

TRICK TEN—Dan led a spade, Art played a heart (trump) and won, Cab played a spade, and Bob played a club (or a diamond).

TRICK ELEVEN—Art led a diamond, Cab played a heart (trump) and won, Dan played a diamond, and Bob played a diamond (or a club).

TRICK TWELVE—Cab led a club, Bob played a heart (trump) and won, Dan played a club, and Art played a club.

TRICK THIRTEEN—Bob led a heart (trump) and won, Cab played a spade, Dan played a spade, and Art played a diamond.

47 / Anthony's Position

The points scored by the three men in the three events may be entered in a three-by-three square array as follows.

	POLE VAULT EVENT	BROAD JUMP EVENT	HIGH JUMP EVENT
Anthony			
Bernard			
Charles			

From [3a] and [3b], the totals of the columns and rows equal the same number. From [2] and [5], let the number of points scored by Anthony and Charles in the broad jump be b points. From [2] and [6], let the number of points scored by Anthony and Bernard in the high jump be h points. From [1] and [2], b may be 0, 1, 2, or 3 and h may be 0, 1, 2, or 3; so there are sixteen possible pairs of values for b and h.

If $b = h$ (if b and h are both 0, 1, 2, or 3), then in order to satisfy [3a] and [3b] the square array becomes as follows.

a	b	b
b	a	b
b	b	a

This situation contradicts [4], so it is impossible.

If $b = 0$ and h does not equal 0 ($b = 0$, $h = 1$; $b = 0$, $h = 2$; $b = 0$, $h = 3$), then in order to satisfy [3a] column two has to have a total equal to that of column three.

________	0	h
________	$2h + a$	h
________	0	a

In order to satisfy [3b] row two has to have a total equal to that of each column. But row two already has a total greater than either column; so this situation is impossible.

If $h = 0$ and b does not equal 0 ($b = 1$, $h = 0$; $b = 2$, $h = 0$; $b = 3$, $h = 0$), then in order to satisfy [3a] column three has to have a total equal to that of column two.

________	b	0
________	a	0
________	b	$2b + a$

In order to satisfy [3b] row three has to have a total equal to that of each column. But row three already has a total greater than either column; so this situation is impossible.

If $b = 1$ and $h = 3$, then in order to satisfy [3a] column two has to have a total equal to that of column three.

________	1	3
________	$a + 4$	3
________	1	a

This situation contradicts [1]: a cannot be less than 0, so $a + 4$ is at least four points. (Also, row two already has a total greater than either column, contradicting [3b].) So this situation is impossible.

If $b = 3$ and $h = 1$, then in order to satisfy [3a] column three has to have a total equal to that of column two.

________	3	1
________	a	1
________	3	$a + 4$

This situation parallels the previous one; so it is impossible.

If $b = 2$ and $h = 3$, then in order to satisfy [3a] column two has to have a total equal to that of column three.

________	2	3
________	$a + 2$	3
________	2	a

In order to satisfy [3b] row three has to have a total equal to that of each column. Then four points would have to be scored by Charles in the pole vault, which contradicts [1]. So this situation is impossible.

If $b = 3$ and $h = 2$, then in order to satisfy [3a] column two has to have a total equal to that of column three.

________	3	2
________	a	2
________	3	$a + 2$

This situation parallels the previous one; so it is impossible.

If $b = 1$ and $h = 2$ or if $b = 2$ and $h = 1$ (the only remaining possibilities), then in order to satisfy [3a] and [3b] the square array becomes either one of the following.

$a + 1$	1	2	*or*	$a + 1$	2	1
0	$a + 2$	2		3	a	1
3	1	a		0	2	$a + 2$

The problem requires finding the value of $a + 1$ (which is the only identical entry in the two arrays): a cannot be greater than 0, otherwise condition [7] would be contradicted; so a must equal 0. Then $a + 1 = 1$.

Since one point is scored for third position, ***Anthony came in third in the pole vault.***

In summary, the points scored were either one of the following:

	POLE VAULT EVENT	BROAD JUMP EVENT	HIGH JUMP EVENT
Anthony	1	1	2
Bernard	0	2	2
Charles	3	1	0

or

	POLE VAULT EVENT	BROAD JUMP EVENT	HIGH JUMP EVENT
Anthony	1	2	1
Bernard	3	0	1
Charles	0	2	2

48 / The Baseball Pennant

From [1], six games were played.

From [2] and the fact that six games were played, one team won three games, one team won two, one team won one, and one team won none (there were no ties). Neither the Sexton team nor the Treble team could have won three games, from the fact that they each scored only one run in a game. The Sexton team did not lose all three games because it scored the highest number of runs (7 runs) in one game. So either (S = Sexton team, T = Treble team, U = Ulster team, and V = Verdue team):

I. V won three games and T won no games,
II. V won three games and U won no games,
III. U won three games and T won no games, or
IV. U won three games and V won no games.

In addition to [1] the following reasoning uses [3] and [4]:

If I is the correct situation, then T scored 6 to S's 7, then T scored 4 against V or U. If V, then [4] could not be satisfied for the third game T played. So the other team was U, and U scored 6. Then T scored 1 to

V's 4, then *V* scored 2 to *S*'s 1, then *V* scored 5 to *U*'s 3, then *S* scored 3 to *U*'s 2. Then these are the scores, arranged to represent rounds:

S T	*T U*	*T V*
7–6	4–6	1–4
V U	*V S*	*S U*
5–3	2–1	3–2

This situation contradicts [5], so it is eliminated.

If II is the correct situation, then *U* scored 6 to *S*'s 7. Then, from [4], *U* scored 2 to *V*'s 5 or *U* scored 3 to *T*'s 6. If the former, then *U* scored 3 to *T*'s 4, then *V* scored 4 to *S*'s 3, then *V* scored 2 to *T*'s 1. But then the last score would be *S*:1 to *T*:6, which contradicts [4]. So *U* scored 3 to *T*'s 6. Then *U* scored 2 to *V*'s 4. Then either *V* scored 2 to *T*'s 1 or *T* scored 2 to *S*'s 1. If the former, then *V* scored 5 to *S*'s 3. But then the last score would be *S*:1 to *T*:4, which contradicts [4]. So *V* scored 2 to *S*'s 1. Then *V* scored 5 to *T*'s 4, then *S* scored 3 to *T*'s 2. Then these are the scores, arranged to represent rounds:

U S	*U T*	*U V*
6–7	3–6	2–4
V T	*V S*	*S T*
5–4	2–1	3–2

This situation contradicts [5], so it is eliminated.

If III is the correct situation, then *T* scored 6 to *S*'s 7. Then *T* scored 1 against *U* or *V*. If *U*, then [4] could not be satisfied for the third game *T* played. So the other team was *V*, and *V* scored 4. Then *T* scored 4 to *U*'s 6, then *U* scored 2 to *S*'s 1, then *U* scored 3 to *V*'s 2, then *S* scored 3 to *V*'s 5. Then these are the scores, arranged to represent rounds:

T S	*T V*	*T U*
6–7	1–4	4–6
U V	*U S*	*S V*
3–2	2–1	3–5

This situation contradicts [5], so it is eliminated.

If IV is the correct situation, then *V* scored 4 to *S*'s 7. Then *V* scored 5 to either *T*'s 6 or *U*'s 6. Then *U* scored 2 to *S*'s 1 or *T*'s 1. So there are four possibilities:

(a)	if	*V S*	*V T*	*U S*	then	*V U*	then	*U T*	then	*S T*
		4–7	5–6	2–1		2–3		6–4		3–1
(b)	if	*V S*	*V T*	*U T*	then	*V U*	then	*U S*	then	*S T*
		4–7	5–6	2–1		2–3		6–3		1–4
									or	
								U S	then	*S T*
								6–1		3–4
(c)	if	*V S*	*V U*	*U S*	then	*V T*	then	*U T*	then	*S T*
		4–7	5–6	2–1		2–4		3–1		3–6
(d)	if	*V S*	*V U*	*U T*	then	*V T*	then	*U S*	then	*S T*
		4–7	5–6	2–1		2–4		3–1		3–6

From [4], (b), (c), and (d) are eliminated. Then these are the scores, arranged to represent rounds:

(a)	*V S*	*V U*	*V T*
	4–7	2–3	5–6
	U T	*S T*	*U S*
	6–4	3–1	2–1

This situation is the only one left and satisfies [5].

From [5] and [6], { *V* T 5–6, *U S* 2–1 } represents the last round.

From [6], the Alleycats are team *V*, the Cougars are team *T*, the Bobcats are team *S*, and the Domestics are team *U*.

Since situation IV is the correct one, *U* won all the games. Since team *U* is the Domestics, ***the Domestics won the pennant.***

49 / No Cause for Celebration

There are four possible boy-girl combinations for the second and third children of each family (*s* stands for sister, *b* for brother):

second child	*b*	*b*	*s*	*s*
third child	*b*	*s*	*b*	*s*

From [1] and [2], these combinations can be developed as follows:

	I	II	III	IV
second child	*b*	*b*	*s*	*s*
third child	*b*	*s*	*b*	*s*
other children	*b* *b* *s* *s*	*b* *b* *b* *s* *s*	*b* *b* *s*	*b* *b* *b* *s*
totals	4*b* 2*s*	4*b* 3*s*	3*b* 2*s*	3*b* 3*s*

From [6] and the above combinations, one family has five children, one family has six children, and one family has seven children. Since only two families can have a sixth child, [5] indicates that the Smith family has five children; therefore, the Smith children are like those in combination III.

The following combinations are left for Brown and Jones:

	BROWN	JONES
(i)	4*b*—3*s*	4*b*—2*s*
(ii)	4*b*—3*s*	3*b*—3*s*
(iii)	4*b*—2*s*	4*b*—3*s*
(iv)	3*b*—3*s*	4*b*—3*s*

From [5], in (i) it is impossible for the number of brothers of one child in the Brown family to equal the number of sisters of one child in the Jones family.

Using the previous combinations for second and third children and referring to [3], [4], and [5], it is possible to begin expanding the com-combinations. It turns out that two variations are possible for each of (ii), (iii), and (iv).

		(ii)			(iii)			(iv)	
	BROWN 4*b*—3*s*	JONES 3*b*—3*s*	SMITH 3*b*—2*s*	BROWN 4*b*—2*s*	JONES 4*b*—3*s*	SMITH 3*b*—2*s*	BROWN 3*b*—3*s*	JONES 4*b*—3*s*	SMITH 3*b*—2*s*
1st									
2nd	*b*	*s*	*s*	*b*	*b*	*s*	*s*	*b*	*s*
3rd	*s*	*s*	*b*	*b*	*s*	*b*	*s*	*s*	*b*
4th		*s*	*s*		*b*	*s*		*b*	*s*
5th	*s*		*b*	*b*		*b*	*s*		*b*
6th	*b*	*b*		*b*	*b*		*b*	*s*	
7th									

1st									
2nd	*b*	*s*	*s*	*b*	*b*	*s*	*s*	*b*	*s*
3rd	*s*	*s*	*b*	*b*	*s*	*b*	*s*	*s*	*b*
4th		*b*	*b*		*b*	*s*		*b*	*s*
5th	*s*		*b*	*s*		*s*	*s*		*b*
6th	*b*	*b*		*b*	*b*		*s*	*b*	
7th									

The second tables for (iii) and (iv) are impossible from the total number of girls allowed for Smith and Brown, respectively.

The first table for (ii) is impossible from the total number of boys allowed for Jones and Smith and from [7].

The first table for (iv) is impossible from the total number of boys allowed for Brown and Smith and from [7].

In the second table for (ii) the seventh child of Brown must be a girl, from [8]. Then, from the totals for Brown, the first child must be a boy. This first born has to be the only boy and this last born has to be the only girl, from [7] and [8] respectively. But these two children could not have married, as required by [9]. So this table is impossible.

The only table left is the first table for (iii). Brown's column and Smith's column can be completed from the totals for Brown and Smith, respectively. Then Jones' column can be completed from [7] and [8] and from the totals for Jones.

	BROWN 4*b*—2*s*	JONES 4*b*—3*s*	SMITH 3*b*—2*s*
1st	*s*	*s*	*b*
2nd	*b*	*b*	*s*
3rd	*b*	*s*	*b*
4th	*s*	*b*	*s*
5th	*b*	*b*	*b*
6th	*b*	*b*	–
7th	–	*s*	–

From [9], a Smith boy married a Jones girl. Therefore, ***the Brown family had no cause for celebration that day.***

50 / The One-Dollar Bill

From [6], a first man must have made an exchange of coins with a second man, and then the second man must have made an exchange of coins with the third man; in these exchanges the first man must have given all of his coins to the second man. Therefore, the amount the first man had must be expressible in two combinations of coins where: different denominations of coins are in each combination (from [6]), each combination of coins cannot change a larger coin (from [2]), and neither combination of coins contains a penny or a silver dollar (from [1]). Searching for coin values to satisfy these three requirements, one finds only two possible values for the first man's original and final holdings: 30¢ consisting of three dimes or one quarter and one nickel, and 55¢ consisting of one quarter and three dimes or one half dollar and one nickel. So the first man's original complete holding and the second man's original partial or complete holding must have been either (N = nickel, D = dime, Q = quarter, and H = half dollar):

	FIRST MAN	SECOND MAN
I	*QDDD*	*HN . . .*
II	*HN*	*QDDD . . .*
III	*DDD*	*QN . . .*
IV	*QN*	*DDD . . .*

From [6], when the third man then made an exchange with the second man he must have given all of his coins to the second man. The third man could not have had one of the four holdings listed for the first man because then the second man would have received from the third man coins that were of the same denomination as those he had given to the first man, contradicting [6]. So the third man must have received from the second man change for some coin he held. Then the second man must have given the third man at least one coin that he received from the first man and at least one coin he held originally; otherwise, either the first or second man could have originally given change for some coin, contradicting [2]. Therefore at least one coin passed through three hands. What was the denomination of this coin?

Since no one had a silver dollar, a half dollar did not pass through three hands. If it was a nickel that passed through three hands, then II or IV represents the exchange between the first man and the second man. But then the second man would have to have had either two more dimes or one more nickel in order to change a larger coin, which contradicts [2]. So a nickel did not pass through three hands. If it was a dime that passed through three hands, then I or III represents the exchange between the first man and the second man. But then the second man would have to have had either two more dimes or one more nickel in order to change a larger coin, which contradicts [2]. So a dime did not pass through three hands. Therefore, it must have been a quarter that passed through three hands.

So I or IV represents the exchange between the first man and the second man. For these holdings, the second man could not have had either one more nickel or two more dimes because [2] would then be contradicted. He could not have held just one more dime because then he would not be able to exchange this dime for a larger coin, contradicting [6]. He could not have had another half dollar in I because then [2] would be contradicted. So there are now three possible situations:

	FIRST MAN	SECOND MAN
I	*QDDD*	*HN Q*
IVa	*QN*	*DDD Q*
IVb	*QN*	*DDD HQ*

Since the quarter that passed through three hands could not have been used to change a silver dollar, from [1], the quarter must have been used to change a half dollar. So after the exchange between the first man and the second man, the third man must have given the second man a half dollar in exchange for at least one quarter. In I and IVb, such an exchange contradicts [6]. So IVa represents the correct holding.

In summary:

	ORIGINAL HOLDING	HOLDING AFTER FIRST EXCHANGE	HOLDING AFTER SECOND EXCHANGE
First man	*QN*	*DDD*	*DDD*
Second man	*QDDD*	*QQN*	*HN*
Third man	*H*	*H*	*QQ*

From [4] and [5], the first man's check must have been for 10¢ or 20¢, the second man's check must have been for 5¢ or 50¢, and the third man's check must have been for 25¢. So one of four sets of checks is correct:

	FIRST MAN	SECOND MAN	THIRD MAN
(i)	20¢	5¢	25¢
(ii)	10¢	5¢	25¢
(iii)	10¢	50¢	25¢
(iv)	20¢	50¢	25¢

Combination (iv) is not possible because, from [1], the woman had at least one coin to begin with (no pennies) and, from [8], after the checks were paid the value of the coins she had totaled less than one dollar.

If combination (i) was the correct one, before the bills were paid the woman did not have any quarters. (The third man had a half dollar at first and his check was for 25¢; but, from [4], she could not change the half dollar.) Also, the woman did not have any nickels (from the second man's original holding and [4]), any half dollars (from [8]), nor any dimes (from the first man's original holding and [4]). So combination (i) contradicts [1] (she had at least one coin); therefore, combination (i) is not possible.

If combination (ii) was the correct one, the woman did not have any nickels (from the second man's original holding and [4]), any quarters (from the third man's original holding and [4]), any half dollars (from the second man's original holding and [4]), nor two or more dimes (from the first man's original holding and [4]). So she had one dime for this combination.

If combination (iii) was the correct one, the woman did not have any nickels (from the second man's original holding and [4]), any quarters (from the third man's original holding and [4]), any half dollars (from [8]), nor two or more dimes (from the first man's original holding and [4]). So she had one dime for this combination.

If combination (ii) was the correct one, after the bills were paid the woman had *QDDN*, the first man had *DD*, the second man had *H*, and the third man had *Q*. From [8], the difference between the value of the coins the woman had and $1 is equal to the cost of the candy. So the candy cost 50¢. But, from [7], the man who bought the candy had a value in coins that exceeded the value of the candy. Then none of the men could have bought the candy. Therefore, combination (ii) is not possible.

Combination (iii) must be the correct one. From [3], Ned is the first man, Lou is the second man, and Moe is the third man. After the bills were paid the woman had *HQDD*, Ned had *DD*, Lou had *N*, and Moe had *Q*. From [8], the candy must have cost 5¢. So, from [7], neither Lou who had the nickel nor Moe who had the quarter bought the candy. Therefore, Ned who had the two dimes must have bought it. So ***Ned gave the woman the one-dollar bill.***

In summary:

	AMOUNT OF CHECK	ORIGINAL HOLDING	HOLDING AFTER FIRST EXCHANGE	HOLDING AFTER SECOND EXCHANGE	HOLDING AFTER CHECKS WERE PAID
Lou	50¢	*QDDD*	*QQN*	*HN*	*N*
Moe	25¢	*H*	*H*	*QQ*	*Q*
Ned	10¢	*QN*	*DDD*	*DDD*	*DD*
Owner	—	*D*	*D*	*D*	*HQDD*

cost of candy: 5¢

A CATALOG OF SELECTED

DOVER BOOKS

IN ALL FIELDS OF INTEREST

A CATALOG OF SELECTED DOVER BOOKS IN ALL FIELDS OF INTEREST

DRAWINGS OF REMBRANDT, edited by Seymour Slive. Updated Lippmann, Hofstede de Groot edition, with definitive scholarly apparatus. All portraits, biblical sketches, landscapes, nudes. Oriental figures, classical studies, together with selection of work by followers. 550 illustrations. Total of 630pp. 9⅛ × 12¼. 21485-0, 21486-9 Pa., Two-vol. set $29.90

GHOST AND HORROR STORIES OF AMBROSE BIERCE, Ambrose Bierce. 24 tales vividly imagined, strangely prophetic, and decades ahead of their time in technical skill: "The Damned Thing," "An Inhabitant of Carcosa," "The Eyes of the Panther," "Moxon's Master," and 20 more. 199pp. 5⅜ × 8½. 20767-6 Pa. $3.95

ETHICAL WRITINGS OF MAIMONIDES, Maimonides. Most significant ethical works of great medieval sage, newly translated for utmost precision, readability. Laws Concerning Character Traits, Eight Chapters, more. 192pp. 5⅜ × 8½. 24522-5 Pa. $4.50

THE EXPLORATION OF THE COLORADO RIVER AND ITS CANYONS, J. W. Powell. Full text of Powell's 1,000-mile expedition down the fabled Colorado in 1869. Superb account of terrain, geology, vegetation, Indians, famine, mutiny, treacherous rapids, mighty canyons, during exploration of last unknown part of continental U.S. 400pp. 5⅜ × 8½. 20094-9 Pa. $7.95

HISTORY OF PHILOSOPHY, Julián Marías. Clearest one-volume history on the market. Every major philosopher and dozens of others, to Existentialism and later. 505pp. 5⅜ × 8½. 21739-6 Pa. $9.95

ALL ABOUT LIGHTNING, Martin A. Uman. Highly readable non-technical survey of nature and causes of lightning, thunderstorms, ball lightning, St. Elmo's Fire, much more. Illustrated. 192pp. 5⅜ × 8½. 25237-X Pa. $5.95

SAILING ALONE AROUND THE WORLD, Captain Joshua Slocum. First man to sail around the world, alone, in small boat. One of great feats of seamanship told in delightful manner. 67 illustrations. 294pp. 5⅜ × 8½. 20326-3 Pa. $4.95

LETTERS AND NOTES ON THE MANNERS, CUSTOMS AND CONDITIONS OF THE NORTH AMERICAN INDIANS, George Catlin. Classic account of life among Plains Indians: ceremonies, hunt, warfare, etc. 312 plates. 572pp. of text. 6⅛ × 9¼. 22118-0, 22119-9, Pa. Two-vol. set $17.90

ALASKA: The Harriman Expedition, 1899, John Burroughs, John Muir, et al. Informative, engrossing accounts of two-month, 9,000-mile expedition. Native peoples, wildlife, forests, geography, salmon industry, glaciers, more. Profusely illustrated. 240 black-and-white line drawings. 124 black-and-white photographs. 3 maps. Index. 576pp. 5⅜ × 8½. 25109-8 Pa. $11.95

ILLUSTRATED DICTIONARY OF HISTORIC ARCHITECTURE, edited by Cyril M. Harris. Extraordinary compendium of clear, concise definitions for over 5,000 important architectural terms complemented by over 2,000 line drawings. Covers full spectrum of architecture from ancient ruins to 20th-century Modernism. Preface. 592pp. 7½ × 9⅝. 24444-X Pa. $15.95

THE NIGHT BEFORE CHRISTMAS, Clement Moore. Full text, and woodcuts from original 1848 book. Also critical, historical material. 19 illustrations. 40pp. 4⅝ × 6. 22797-9 Pa. $2.50

THE LESSON OF JAPANESE ARCHITECTURE: 165 Photographs, Jiro Harada. Memorable gallery of 165 photographs taken in the 1930's of exquisite Japanese homes of the well-to-do and historic buildings. 13 line diagrams. 192pp. 8⅜ × 11¼. 24778-3 Pa. $10.95

THE AUTOBIOGRAPHY OF CHARLES DARWIN AND SELECTED LETTERS, edited by Francis Darwin. The fascinating life of eccentric genius composed of an intimate memoir by Darwin (intended for his children); commentary by his son, Francis; hundreds of fragments from notebooks, journals, papers; and letters to and from Lyell, Hooker, Huxley, Wallace and Henslow. xi + 365pp. 5⅜ × 8. 20479-0 Pa. $6.95

WONDERS OF THE SKY: Observing Rainbows, Comets, Eclipses, the Stars and Other Phenomena, Fred Schaaf. Charming, easy-to-read poetic guide to all manner of celestial events visible to the naked eye. Mock suns, glories, Belt of Venus, more. Illustrated. 299pp. 5¼ × 8¼. 24402-4 Pa. $7.95

BURNHAM'S CELESTIAL HANDBOOK, Robert Burnham, Jr. Thorough guide to the stars beyond our solar system. Exhaustive treatment. Alphabetical by constellation: Andromeda to Cetus in Vol. 1; Chamaeleon to Orion in Vol. 2; and Pavo to Vulpecula in Vol. 3. Hundreds of illustrations. Index in Vol. 3. 2,000pp. 6⅛ × 9¼. 23567-X, 23568-8, 23673-0 Pa., Three-vol. set $41.85

STAR NAMES: Their Lore and Meaning, Richard Hinckley Allen. Fascinating history of names various cultures have given to constellations and literary and folkloristic uses that have been made of stars. Indexes to subjects. Arabic and Greek names. Biblical references. Bibliography. 563pp. 5⅜ × 8½. 21079-0 Pa. $8.95

THIRTY YEARS THAT SHOOK PHYSICS: The Story of Quantum Theory, George Gamow. Lucid, accessible introduction to influential theory of energy and matter. Careful explanations of Dirac's anti-particles, Bohr's model of the atom, much more. 12 plates. Numerous drawings. 240pp. 5⅜ × 8½. 24895-X Pa. $5.95

CHINESE DOMESTIC FURNITURE IN PHOTOGRAPHS AND MEASURED DRAWINGS, Gustav Ecke. A rare volume, now affordably priced for antique collectors, furniture buffs and art historians. Detailed review of styles ranging from early Shang to late Ming. Unabridged republication. 161 black-and-white drawings, photos. Total of 224pp. 8⅜ × 11¼. (Available in U.S. only) 25171-3 Pa. $13.95

VINCENT VAN GOGH: A Biography, Julius Meier-Graefe. Dynamic, penetrating study of artist's life, relationship with brother, Theo, painting techniques, travels, more. Readable, engrossing. 160pp. 5⅜ × 8½. (Available in U.S. only) 25253-1 Pa. $4.95

HOW TO WRITE, Gertrude Stein. Gertrude Stein claimed anyone could understand her unconventional writing—here are clues to help. Fascinating improvisations, language experiments, explanations illuminate Stein's craft and the art of writing. Total of 414pp. 4⅝ × 6⅜. 23144-5 Pa. $6.95

ADVENTURES AT SEA IN THE GREAT AGE OF SAIL: Five Firsthand Narratives, edited by Elliot Snow. Rare true accounts of exploration, whaling, shipwreck, fierce natives, trade, shipboard life, more. 33 illustrations. Introduction. 353pp. 5⅜ × 8½. 25177-2 Pa. $8.95

THE HERBAL OR GENERAL HISTORY OF PLANTS, John Gerard. Classic descriptions of about 2,850 plants—with over 2,700 illustrations—includes Latin and English names, physical descriptions, varieties, time and place of growth, more. 2,706 illustrations. xlv + 1,678pp. 8½ × 12¼. 23147-X Cloth. $75.00

DOROTHY AND THE WIZARD IN OZ, L. Frank Baum. Dorothy and the Wizard visit the center of the Earth, where people are vegetables, glass houses grow and Oz characters reappear. Classic sequel to *Wizard of Oz*. 256pp. 5⅜ × 8. 24714-7 Pa. $5.95

SONGS OF EXPERIENCE: Facsimile Reproduction with 26 Plates in Full Color, William Blake. This facsimile of Blake's original "Illuminated Book" reproduces 26 full-color plates from a rare 1826 edition. Includes "The Tyger," "London," "Holy Thursday," and other immortal poems. 26 color plates. Printed text of poems. 48pp. 5¼ × 7. 24636-1 Pa. $3.50

SONGS OF INNOCENCE, William Blake. The first and most popular of Blake's famous "Illuminated Books," in a facsimile edition reproducing all 31 brightly colored plates. Additional printed text of each poem. 64pp. 5¼ × 7. 22764-2 Pa. $3.50

PRECIOUS STONES, Max Bauer. Classic, thorough study of diamonds, rubies, emeralds, garnets, etc.: physical character, occurrence, properties, use, similar topics. 20 plates, 8 in color. 94 figures. 659pp. 6⅛ × 9¼. 21910-0, 21911-9 Pa., Two-vol. set $15.90

ENCYCLOPEDIA OF VICTORIAN NEEDLEWORK, S. F. A. Caulfeild and Blanche Saward. Full, precise descriptions of stitches, techniques for dozens of needlecrafts—most exhaustive reference of its kind. Over 800 figures. Total of 679pp. 8⅛ × 11. Two volumes. Vol. 1 22800-2 Pa. $11.95
Vol. 2 22801-0 Pa. $11.95

THE MARVELOUS LAND OF OZ, L. Frank Baum. Second Oz book, the Scarecrow and Tin Woodman are back with hero named Tip, Oz magic. 136 illustrations. 287pp. 5⅜ × 8½. 20692-0 Pa. $5.95

WILD FOWL DECOYS, Joel Barber. Basic book on the subject, by foremost authority and collector. Reveals history of decoy making and rigging, place in American culture, different kinds of decoys, how to make them, and how to use them. 140 plates. 156pp. 7⅞ × 10¾. 20011-6 Pa. $8.95

HISTORY OF LACE, Mrs. Bury Palliser. Definitive, profusely illustrated chronicle of lace from earliest times to late 19th century. Laces of Italy, Greece, England, France, Belgium, etc. Landmark of needlework scholarship. 266 illustrations. 672pp. 6⅛ × 9¼. 24742-2 Pa. $14.95

ILLUSTRATED GUIDE TO SHAKER FURNITURE, Robert Meader. All furniture and appurtenances, with much on unknown local styles. 235 photos. 146pp. 9 × 12. 22819-3 Pa. $8.95

WHALE SHIPS AND WHALING: A Pictorial Survey, George Francis Dow. Over 200 vintage engravings, drawings, photographs of barks, brigs, cutters, other vessels. Also harpoons, lances, whaling guns, many other artifacts. Comprehensive text by foremost authority. 207 black-and-white illustrations. 288pp. 6 × 9. 24808-9 Pa. $8.95

THE BERTRAMS, Anthony Trollope. Powerful portrayal of blind self-will and thwarted ambition includes one of Trollope's most heartrending love stories. 497pp. 5⅜ × 8½. 25119-5 Pa. $9.95

ADVENTURES WITH A HAND LENS, Richard Headstrom. Clearly written guide to observing and studying flowers and grasses, fish scales, moth and insect wings, egg cases, buds, feathers, seeds, leaf scars, moss, molds, ferns, common crystals, etc.—all with an ordinary, inexpensive magnifying glass. 209 exact line drawings aid in your discoveries. 220pp. 5⅜ × 8½. 23330-8 Pa. $4.95

RODIN ON ART AND ARTISTS, Auguste Rodin. Great sculptor's candid, wide-ranging comments on meaning of art; great artists; relation of sculpture to poetry, painting, music; philosophy of life, more. 76 superb black-and-white illustrations of Rodin's sculpture, drawings and prints. 119pp. 8⅝ × 11¼. 24487-3 Pa. $7.95

FIFTY CLASSIC FRENCH FILMS, 1912–1982: A Pictorial Record, Anthony Slide. Memorable stills from Grand Illusion, Beauty and the Beast, Hiroshima, Mon Amour, many more. Credits, plot synopses, reviews, etc. 160pp. 8¼ × 11. 25256-6 Pa. $11.95

THE PRINCIPLES OF PSYCHOLOGY, William James. Famous long course complete, unabridged. Stream of thought, time perception, memory, experimental methods; great work decades ahead of its time. 94 figures. 1,391pp. 5⅜ × 8½. 20381-6, 20382-4 Pa., Two-vol. set $23.90

BODIES IN A BOOKSHOP, R. T. Campbell. Challenging mystery of blackmail and murder with ingenious plot and superbly drawn characters. In the best tradition of British suspense fiction. 192pp. 5⅜ × 8½. 24720-1 Pa. $3.95

CALLAS: PORTRAIT OF A PRIMA DONNA, George Jellinek. Renowned commentator on the musical scene chronicles incredible career and life of the most controversial, fascinating, influential operatic personality of our time. 64 black-and-white photographs. 416pp. 5⅜ × 8¼. 25047-4 Pa. $8.95

GEOMETRY, RELATIVITY AND THE FOURTH DIMENSION, Rudolph Rucker. Exposition of fourth dimension, concepts of relativity as Flatland characters continue adventures. Popular, easily followed yet accurate, profound. 141 illustrations. 133pp. 5⅜ × 8½. 23400-2 Pa. $4.95

HOUSEHOLD STORIES BY THE BROTHERS GRIMM, with pictures by Walter Crane. 53 classic stories—Rumpelstiltskin, Rapunzel, Hansel and Gretel, the Fisherman and his Wife, Snow White, Tom Thumb, Sleeping Beauty, Cinderella, and so much more—lavishly illustrated with original 19th century drawings. 114 illustrations. x + 269pp. 5⅜ × 8½. 21080-4 Pa. $4.95

SUNDIALS, Albert Waugh. Far and away the best, most thorough coverage of ideas, mathematics concerned, types, construction, adjusting anywhere. Over 100 illustrations. 230pp. 5⅜ × 8½. 22947-5 Pa. $4.95

PICTURE HISTORY OF THE NORMANDIE: With 190 Illustrations, Frank O. Braynard. Full story of legendary French ocean liner: Art Deco interiors, design innovations, furnishings, celebrities, maiden voyage, tragic fire, much more. Extensive text. 144pp. 8⅞ × 11¾. 25257-4 Pa. $10.95

THE FIRST AMERICAN COOKBOOK: A Facsimile of "American Cookery," 1796, Amelia Simmons. Facsimile of the first American-written cookbook published in the United States contains authentic recipes for colonial favorites—pumpkin pudding, winter squash pudding, spruce beer, Indian slapjacks, and more. Introductory Essay and Glossary of colonial cooking terms. 80pp. 5⅜ × 8½. 24710-4 Pa. $3.50

101 PUZZLES IN THOUGHT AND LOGIC, C. R. Wylie, Jr. Solve murders and robberies, find out which fishermen are liars, how a blind man could possibly identify a color—purely by your own reasoning! 107pp. 5⅜ × 8½. 20367-0 Pa. $2.50

THE BOOK OF WORLD-FAMOUS MUSIC—CLASSICAL, POPULAR AND FOLK, James J. Fuld. Revised and enlarged republication of landmark work in musico-bibliography. Full information about nearly 1,000 songs and compositions including first lines of music and lyrics. New supplement. Index. 800pp. 5⅜ × 8¼. 24857-7 Pa. $15.95

ANTHROPOLOGY AND MODERN LIFE, Franz Boas. Great anthropologist's classic treatise on race and culture. Introduction by Ruth Bunzel. Only inexpensive paperback edition. 255pp. 5⅜ × 8½. 25245-0 Pa. $6.95

THE TALE OF PETER RABBIT, Beatrix Potter. The inimitable Peter's terrifying adventure in Mr. McGregor's garden, with all 27 wonderful, full-color Potter illustrations. 55pp. 4¼ × 5½. (Available in U.S. only) 22827-4 Pa. $1.75

THREE PROPHETIC SCIENCE FICTION NOVELS, H. G. Wells. *When the Sleeper Wakes, A Story of the Days to Come* and *The Time Machine* (full version). 335pp. 5⅜ × 8½. (Available in U.S. only) 20605-X Pa. $6.95

APICIUS COOKERY AND DINING IN IMPERIAL ROME, edited and translated by Joseph Dommers Vehling. Oldest known cookbook in existence offers readers a clear picture of what foods Romans ate, how they prepared them, etc. 49 illustrations. 301pp. 6⅛ × 9¼. 23563-7 Pa. $7.95

SHAKESPEARE LEXICON AND QUOTATION DICTIONARY, Alexander Schmidt. Full definitions, locations, shades of meaning of every word in plays and poems. More than 50,000 exact quotations. 1,485pp. 6½ × 9¼.
22726-X, 22727-8 Pa., Two-vol. set $29.90

THE WORLD'S GREAT SPEECHES, edited by Lewis Copeland and Lawrence W. Lamm. Vast collection of 278 speeches from Greeks to 1970. Powerful and effective models; unique look at history. 842pp. 5⅜ × 8½. 20468-5 Pa. $11.95

PLANTS OF THE BIBLE, Harold N. Moldenke and Alma L. Moldenke. Standard reference to all 230 plants mentioned in Scriptures. Latin name, biblical reference, uses, modern identity, much more. Unsurpassed encyclopedic resource for scholars, botanists, nature lovers, students of Bible. Bibliography. Indexes. 123 black-and-white illustrations. 384pp. 6 × 9. 25069-5 Pa. $8.95

FAMOUS AMERICAN WOMEN: A Biographical Dictionary from Colonial Times to the Present, Robert McHenry, ed. From Pocahontas to Rosa Parks, 1,035 distinguished American women documented in separate biographical entries. Accurate, up-to-date data, numerous categories, spans 400 years. Indices. 493pp. 6½ × 9¼. 24523-3 Pa. $10.95

THE FABULOUS INTERIORS OF THE GREAT OCEAN LINERS IN HISTORIC PHOTOGRAPHS, William H. Miller, Jr. Some 200 superb photographs capture exquisite interiors of world's great "floating palaces"—1890's to 1980's: *Titanic, Ile de France, Queen Elizabeth, United States, Europa,* more. Approx. 200 black-and-white photographs. Captions. Text. Introduction. 160pp. 8⅞ × 11¾. 24756-2 Pa. $9.95

THE GREAT LUXURY LINERS, 1927–1954: A Photographic Record, William H. Miller, Jr. Nostalgic tribute to heyday of ocean liners. 186 photos of Ile de France, Normandie, Leviathan, Queen Elizabeth, United States, many others. Interior and exterior views. Introduction. Captions. 160pp. 9 × 12. 24056-8 Pa. $10.95

A NATURAL HISTORY OF THE DUCKS, John Charles Phillips. Great landmark of ornithology offers complete detailed coverage of nearly 200 species and subspecies of ducks: gadwall, sheldrake, merganser, pintail, many more. 74 full-color plates, 102 black-and-white. Bibliography. Total of 1,920pp. 8⅝ × 11¼. 25141-1, 25142-X Cloth. Two-vol. set $100.00

THE SEAWEED HANDBOOK: An Illustrated Guide to Seaweeds from North Carolina to Canada, Thomas F. Lee. Concise reference covers 78 species. Scientific and common names, habitat, distribution, more. Finding keys for easy identification. 224pp. 5⅜ × 8½. 25215-9 Pa. $6.95

THE TEN BOOKS OF ARCHITECTURE: The 1755 Leoni Edition, Leon Battista Alberti. Rare classic helped introduce the glories of ancient architecture to the Renaissance. 68 black-and-white plates. 336pp. 8⅜ × 11¼. 25239-6 Pa. $14.95

MISS MACKENZIE, Anthony Trollope. Minor masterpieces by Victorian master unmasks many truths about life in 19th-century England. First inexpensive edition in years. 392pp. 5⅜ × 8½. 25201-9 Pa. $8.95

THE RIME OF THE ANCIENT MARINER, Gustave Doré, Samuel Taylor Coleridge. Dramatic engravings considered by many to be his greatest work. The terrifying space of the open sea, the storms and whirlpools of an unknown ocean, the ice of Antarctica, more—all rendered in a powerful, chilling manner. Full text. 38 plates. 77pp. 9¼ × 12. 22305-1 Pa. $4.95

THE EXPEDITIONS OF ZEBULON MONTGOMERY PIKE, Zebulon Montgomery Pike. Fascinating first-hand accounts (1805–6) of exploration of Mississippi River, Indian wars, capture by Spanish dragoons, much more. 1,088pp. 5⅜ × 8½. 25254-X, 25255-8 Pa. Two-vol. set $25.90

A CONCISE HISTORY OF PHOTOGRAPHY: Third Revised Edition, Helmut Gernsheim. Best one-volume history—camera obscura, photochemistry, daguerreotypes, evolution of cameras, film, more. Also artistic aspects—landscape, portraits, fine art, etc. 281 black-and-white photographs. 26 in color. 176pp. 8⅜ × 11¼. 25128-4 Pa. $13.95

THE DORÉ BIBLE ILLUSTRATIONS, Gustave Doré. 241 detailed plates from the Bible: the Creation scenes, Adam and Eve, Flood, Babylon, battle sequences, life of Jesus, etc. Each plate is accompanied by the verses from the King James version of the Bible. 241pp. 9 × 12. 23004-X Pa. $9.95

HUGGER-MUGGER IN THE LOUVRE, Elliot Paul. Second Homer Evans mystery-comedy. Theft at the Louvre involves sleuth in hilarious, madcap caper. "A knockout."—Books. 336pp. 5⅜ × 8½. 25185-3 Pa. $5.95

FLATLAND, E. A. Abbott. Intriguing and enormously popular science-fiction classic explores the complexities of trying to survive as a two-dimensional being in a three-dimensional world. Amusingly illustrated by the author. 16 illustrations. 103pp. 5⅜ × 8½. 20001-9 Pa. $2.50

THE HISTORY OF THE LEWIS AND CLARK EXPEDITION, Meriwether Lewis and William Clark, edited by Elliott Coues. Classic edition of Lewis and Clark's day-by-day journals that later became the basis for U.S. claims to Oregon and the West. Accurate and invaluable geographical, botanical, biological, meteorological and anthropological material. Total of 1,508pp. 5⅜ × 8½.
21268-8, 21269-6, 21270-X Pa. Three-vol. set $26.85

LANGUAGE, TRUTH AND LOGIC, Alfred J. Ayer. Famous, clear introduction to Vienna, Cambridge schools of Logical Positivism. Role of philosophy, elimination of metaphysics, nature of analysis, etc. 160pp. 5⅜ × 8½. (Available in U.S. and Canada only) 20010-8 Pa. $3.95

MATHEMATICS FOR THE NONMATHEMATICIAN, Morris Kline. Detailed, college-level treatment of mathematics in cultural and historical context, with numerous exercises. For liberal arts students. Preface. Recommended Reading Lists. Tables. Index. Numerous black-and-white figures. xvi + 641pp. 5⅜ × 8½.
24823-2 Pa. $11.95

HANDBOOK OF PICTORIAL SYMBOLS, Rudolph Modley. 3,250 signs and symbols, many systems in full; official or heavy commercial use. Arranged by subject. Most in Pictorial Archive series. 143pp. 8⅛ × 11. 23357-X Pa. $6.95

INCIDENTS OF TRAVEL IN YUCATAN, John L. Stephens. Classic (1843) exploration of jungles of Yucatan, looking for evidences of Maya civilization. Travel adventures, Mexican and Indian culture, etc. Total of 669pp. 5⅜ × 8½.
20926-1, 20927-X Pa., Two-vol. set $11.90

DEGAS: An Intimate Portrait, Ambroise Vollard. Charming, anecdotal memoir by famous art dealer of one of the greatest 19th-century French painters. 14 black-and-white illustrations. Introduction by Harold L. Van Doren. 96pp. 5⅜ × 8½.
25131-4 Pa. $4.95

PERSONAL NARRATIVE OF A PILGRIMAGE TO ALMANDINAH AND MECCAH, Richard Burton. Great travel classic by remarkably colorful personality. Burton, disguised as a Moroccan, visited sacred shrines of Islam, narrowly escaping death. 47 illustrations. 959pp. 5⅜ × 8½. 21217-3, 21218-1 Pa., Two-vol. set $19.90

PHRASE AND WORD ORIGINS, A. H. Holt. Entertaining, reliable, modern study of more than 1,200 colorful words, phrases, origins and histories. Much unexpected information. 254pp. 5⅜ × 8½. 20758-7 Pa. $5.95

THE RED THUMB MARK, R. Austin Freeman. In this first Dr. Thorndyke case, the great scientific detective draws fascinating conclusions from the nature of a single fingerprint. Exciting story, authentic science. 320pp. 5⅜ × 8½. (Available in U.S. only) 25210-8 Pa. $6.95

AN EGYPTIAN HIEROGLYPHIC DICTIONARY, E. A. Wallis Budge. Monumental work containing about 25,000 words or terms that occur in texts ranging from 3000 B.C. to 600 A.D. Each entry consists of a transliteration of the word, the word in hieroglyphs, and the meaning in English. 1,314pp. 6⅞ × 10.
23615-3, 23616-1 Pa., Two-vol. set $31.90

THE COMPLEAT STRATEGYST: Being a Primer on the Theory of Games of Strategy, J. D. Williams. Highly entertaining classic describes, with many illustrated examples, how to select best strategies in conflict situations. Prefaces. Appendices. xvi + 268pp. 5⅜ × 8½. 25101-2 Pa. $5.95

THE ROAD TO OZ, L. Frank Baum. Dorothy meets the Shaggy Man, little Button-Bright and the Rainbow's beautiful daughter in this delightful trip to the magical Land of Oz. 272pp. 5⅜ × 8. 25208-6 Pa. $5.95

POINT AND LINE TO PLANE, Wassily Kandinsky. Seminal exposition of role of point, line, other elements in non-objective painting. Essential to understanding 20th-century art. 127 illustrations. 192pp. 6½ × 9¼. 23808-3 Pa. $5.95

LADY ANNA, Anthony Trollope. Moving chronicle of Countess Lovel's bitter struggle to win for herself and daughter Anna their rightful rank and fortune—perhaps at cost of sanity itself. 384pp. 5⅜ × 8½. 24669-8 Pa. $8.95

EGYPTIAN MAGIC, E. A. Wallis Budge. Sums up all that is known about magic in Ancient Egypt: the role of magic in controlling the gods, powerful amulets that warded off evil spirits, scarabs of immortality, use of wax images, formulas and spells, the secret name, much more. 253pp. 5⅜ × 8½. 22681-6 Pa. $4.50

THE DANCE OF SIVA, Ananda Coomaraswamy. Preeminent authority unfolds the vast metaphysic of India: the revelation of her art, conception of the universe, social organization, etc. 27 reproductions of art masterpieces. 192pp. 5⅜ × 8½.
24817-8 Pa. $5.95

CHRISTMAS CUSTOMS AND TRADITIONS, Clement A. Miles. Origin, evolution, significance of religious, secular practices. Caroling, gifts, yule logs, much more. Full, scholarly yet fascinating; non-sectarian. 400pp. 5⅜ × 8½.
23354-5 Pa. $6.95

THE HUMAN FIGURE IN MOTION, Eadweard Muybridge. More than 4,500 stopped-action photos, in action series, showing undraped men, women, children jumping, lying down, throwing, sitting, wrestling, carrying, etc. 390pp. 7⅞ × 10⅝.
20204-6 Cloth. $21.95

THE MAN WHO WAS THURSDAY, Gilbert Keith Chesterton. Witty, fast-paced novel about a club of anarchists in turn-of-the-century London. Brilliant social, religious, philosophical speculations. 128pp. 5⅜ × 8½. 25121-7 Pa. $3.95

A CEZANNE SKETCHBOOK: Figures, Portraits, Landscapes and Still Lifes, Paul Cezanne. Great artist experiments with tonal effects, light, mass, other qualities in over 100 drawings. A revealing view of developing master painter, precursor of Cubism. 102 black-and-white illustrations. 144pp. 8⅜ × 6⅜. 24790-2 Pa. $5.95

AN ENCYCLOPEDIA OF BATTLES: Accounts of Over 1,560 Battles from 1479 B.C. to the Present, David Eggenberger. Presents essential details of every major battle in recorded history, from the first battle of Megiddo in 1479 B.C. to Grenada in 1984. List of Battle Maps. New Appendix covering the years 1967–1984. Index. 99 illustrations. 544pp. 6½ × 9¼. 24913-1 Pa. $14.95

AN ETYMOLOGICAL DICTIONARY OF MODERN ENGLISH, Ernest Weekley. Richest, fullest work, by foremost British lexicographer. Detailed word histories. Inexhaustible. Total of 856pp. 6½ × 9¼.
21873-2, 21874-0 Pa., Two-vol. set $17.00

WEBSTER'S AMERICAN MILITARY BIOGRAPHIES, edited by Robert McHenry. Over 1,000 figures who shaped 3 centuries of American military history. Detailed biographies of Nathan Hale, Douglas MacArthur, Mary Hallaren, others. Chronologies of engagements, more. Introduction. Addenda. 1,033 entries in alphabetical order. xi + 548pp. 6½ × 9¼. (Available in U.S. only)
24758-9 Pa. $13.95

LIFE IN ANCIENT EGYPT, Adolf Erman. Detailed older account, with much not in more recent books: domestic life, religion, magic, medicine, commerce, and whatever else needed for complete picture. Many illustrations. 597pp. 5⅜ × 8½.
22632-8 Pa. $8.95

HISTORIC COSTUME IN PICTURES, Braun & Schneider. Over 1,450 costumed figures shown, covering a wide variety of peoples: kings, emperors, nobles, priests, servants, soldiers, scholars, townsfolk, peasants, merchants, courtiers, cavaliers, and more. 256pp. 8⅜ × 11¼. 23150-X Pa. $9.95

THE NOTEBOOKS OF LEONARDO DA VINCI, edited by J. P. Richter. Extracts from manuscripts reveal great genius; on painting, sculpture, anatomy, sciences, geography, etc. Both Italian and English. 186 ms. pages reproduced, plus 500 additional drawings, including studies for *Last Supper, Sforza* monument, etc. 860pp. 7⅞ × 10¾. (Available in U.S. only) 22572-0, 22573-9 Pa., Two-vol. set $31.90

CATALOG OF DOVER BOOKS

THE ART NOUVEAU STYLE BOOK OF ALPHONSE MUCHA: All 72 Plates from "Documents Decoratifs" in Original Color, Alphonse Mucha. Rare copyright-free design portfolio by high priest of Art Nouveau. Jewelry, wallpaper, stained glass, furniture, figure studies, plant and animal motifs, etc. Only complete one-volume edition. 80pp. 9⅜ × 12¼. 24044-4 Pa. $9.95

ANIMALS: 1,419 COPYRIGHT-FREE ILLUSTRATIONS OF MAMMALS, BIRDS, FISH, INSECTS, ETC., edited by Jim Harter. Clear wood engravings present, in extremely lifelike poses, over 1,000 species of animals. One of the most extensive pictorial sourcebooks of its kind. Captions. Index. 284pp. 9 × 12. 23766-4 Pa. $9.95

OBELISTS FLY HIGH, C. Daly King. Masterpiece of American detective fiction, long out of print, involves murder on a 1935 transcontinental flight—"a very thrilling story"—NY Times. Unabridged and unaltered republication of the edition published by William Collins Sons & Co. Ltd., London, 1935. 288pp. 5⅜ × 8½. (Available in U.S. only) 25036-9 Pa. $5.95

VICTORIAN AND EDWARDIAN FASHION: A Photographic Survey, Alison Gernsheim. First fashion history completely illustrated by contemporary photographs. Full text plus 235 photos, 1840–1914, in which many celebrities appear. 240pp. 6½ × 9¼. 24205-6 Pa. $6.95

THE ART OF THE FRENCH ILLUSTRATED BOOK, 1700–1914, Gordon N. Ray. Over 630 superb book illustrations by Fragonard, Delacroix, Daumier, Doré, Grandville, Manet, Mucha, Steinlen, Toulouse-Lautrec and many others. Preface. Introduction. 633 halftones. Indices of artists, authors & titles, binders and provenances. Appendices. Bibliography. 608pp. 8⅜ × 11¼. 25086-5 Pa. $24.95

THE WONDERFUL WIZARD OF OZ, L. Frank Baum. Facsimile in full color of America's finest children's classic. 143 illustrations by W. W. Denslow. 267pp. 5⅜ × 8½. 20691-2 Pa. $7.95

FRONTIERS OF MODERN PHYSICS: New Perspectives on Cosmology, Relativity, Black Holes and Extraterrestrial Intelligence, Tony Rothman, et al. For the intelligent layman. Subjects include: cosmological models of the universe; black holes; the neutrino; the search for extraterrestrial intelligence. Introduction. 46 black-and-white illustrations. 192pp. 5⅜ × 8½. 24587-X Pa. $7.95

THE FRIENDLY STARS, Martha Evans Martin & Donald Howard Menzel. Classic text marshalls the stars together in an engaging, non-technical survey, presenting them as sources of beauty in night sky. 23 illustrations. Foreword. 2 star charts. Index. 147pp. 5⅜ × 8½. 21099-5 Pa. $3.95

FADS AND FALLACIES IN THE NAME OF SCIENCE, Martin Gardner. Fair, witty appraisal of cranks, quacks, and quackeries of science and pseudoscience: hollow earth, Velikovsky, orgone energy, Dianetics, flying saucers, Bridey Murphy, food and medical fads, etc. Revised, expanded In the Name of Science. "A very able and even-tempered presentation."—The New Yorker. 363pp. 5⅜ × 8. 20394-8 Pa. $6.95

ANCIENT EGYPT: ITS CULTURE AND HISTORY, J. E Manchip White. From pre-dynastics through Ptolemies: society, history, political structure, religion, daily life, literature, cultural heritage. 48 plates. 217pp. 5⅜ × 8½. 22548-8 Pa. $5.95

SIR HARRY HOTSPUR OF HUMBLETHWAITE, Anthony Trollope. Incisive, unconventional psychological study of a conflict between a wealthy baronet, his idealistic daughter, and their scapegrace cousin. The 1870 novel in its first inexpensive edition in years. 250pp. 5⅜ × 8½. 24953-0 Pa. $5.95

LASERS AND HOLOGRAPHY, Winston E. Kock. Sound introduction to burgeoning field, expanded (1981) for second edition. Wave patterns, coherence, lasers, diffraction, zone plates, properties of holograms, recent advances. 84 illustrations. 160pp. 5⅜ × 8¼. (Except in United Kingdom) 24041-X Pa. $3.95

INTRODUCTION TO ARTIFICIAL INTELLIGENCE: SECOND, ENLARGED EDITION, Philip C. Jackson, Jr. Comprehensive survey of artificial intelligence—the study of how machines (computers) can be made to act intelligently. Includes introductory and advanced material. Extensive notes updating the main text. 132 black-and-white illustrations. 512pp. 5⅜ × 8½. 24864-X Pa. $8.95

HISTORY OF INDIAN AND INDONESIAN ART, Ananda K. Coomaraswamy. Over 400 illustrations illuminate classic study of Indian art from earliest Harappa finds to early 20th century. Provides philosophical, religious and social insights. 304pp. 6⅝ × 9⅝. 25005-9 Pa. $9.95

THE GOLEM, Gustav Meyrink. Most famous supernatural novel in modern European literature, set in Ghetto of Old Prague around 1890. Compelling story of mystical experiences, strange transformations, profound terror. 13 black-and-white illustrations. 224pp. 5⅜ × 8½. (Available in U.S. only) 25025-3 Pa. $6.95

PICTORIAL ENCYCLOPEDIA OF HISTORIC ARCHITECTURAL PLANS, DETAILS AND ELEMENTS: With 1,880 Line Drawings of Arches, Domes, Doorways, Facades, Gables, Windows, etc., John Theodore Haneman. Sourcebook of inspiration for architects, designers, others. Bibliography. Captions. 141pp. 9 × 12. 24605-1 Pa. $7.95

BENCHLEY LOST AND FOUND, Robert Benchley. Finest humor from early 30's, about pet peeves, child psychologists, post office and others. Mostly unavailable elsewhere. 73 illustrations by Peter Arno and others. 183pp. 5⅜ × 8½. 22410-4 Pa. $4.95

ERTÉ GRAPHICS, Erté. Collection of striking color graphics: *Seasons, Alphabet, Numerals, Aces* and *Precious Stones*. 50 plates, including 4 on covers. 48pp. 9⅜ × 12¼. 23580-7 Pa. $7.95

THE JOURNAL OF HENRY D. THOREAU, edited by Bradford Torrey, F. H. Allen. Complete reprinting of 14 volumes, 1837–61, over two million words; the sourcebooks for *Walden*, etc. Definitive. All original sketches, plus 75 photographs. 1,804pp. 8½ × 12¼. 20312-3, 20313-1 Cloth., Two-vol. set $120.00

CASTLES: THEIR CONSTRUCTION AND HISTORY, Sidney Toy. Traces castle development from ancient roots. Nearly 200 photographs and drawings illustrate moats, keeps, baileys, many other features. Caernarvon, Dover Castles, Hadrian's Wall, Tower of London, dozens more. 256pp. 5⅜ × 8¼. 24898-4 Pa. $6.95